The Dual Light-Speed Universe

and the

Dot-Wave Theory

with

Quantum Entanglement

Gerald Grushow

The Dual Light-Speed Universe and the Dot-Wave Theory with Quantum Entanglement by Gerald Grushow

Self-published via CreateSpace Independent Publishing Platform, North Charleston, SC

Available from Amazon.com, CreateSpace.com, and other retail outlets.

Printed by CreateSpace, An Amazon.com Company

ISBN-13: 978-1723217395

ISBN-10: 1723217395

Books by Gerald Grushow

The Natural God of Law, Love, and Truth 1994

Doppler Space Time 2000

Science of God 2001

Aliens Within Us 2005

Futureoids 2013*

Cosmic Reincarnation 2015*

Cosmology of God and the Universe 2015*

Gravity and the Dot-Wave Theory 2015*

The Fabric of the Soul 2015*

Relativity and the Dot-Wave Theory 2016*

The Gravitational Wave and the Dot-Wave Theory 2016*

Futureoids & Cosmic Reincarnation 2018*

The Dual Light-Speed Universe and the Dot-Wave Theory 2018*

Futureoids and the Dual Light-Speed Universe 2018*

Cosmic Reincarnation and the Dual Light-Speed Universe 2018*

Books are available on Amazon.com and other retail outlets. * Also available on kindle.

With my greatest appreciation to my daughter Melissa who has helped with the computer processing, editing, proof reading, and the book cover. Without her valuable ideas and help, this book would not have been possible.

Preface

The purpose of this book is to provide mankind with an understanding of the Dual Light-Speed universe from an Engineering perspective. This will enable the average reader with reasonable math and science skills to understand the complexities of the Dual Light-Speed Universe that we live in.

In general an understanding of the universe has required the greatest mathematical and scientific minds. Einstein's general relativity equations are beyond the ability of most people to grasp. The same is true of the Quantum theory mathematicians. For those who look at String theory, things are so complex that few people could understand what the scientists are talking about.

Since early scientific man, the scientists have attempted to present a picture of the universe with a single light speed Co. Einstein formulated his special relativity and general relativity based upon a single light speed universe. This has been disputed by quantum physicists who specify that the universe operates at a much higher light speed some of the time. The quantum entanglement experiments have shown interactions at a distance at speeds greater than 10,000 Co.

String theory specifies that the universe that we live in is a particular dimension of a higher greater dimensions universe. The net result in the theory presented here is that the universe is the product of a higher light speed universe which compressed toward a pinpoint to produce a dual light speed universe.

The fundamental sub-particle is the dot-wave which oscillates from light speed Co to light speed Cs.

The mathematics of the universe that we see and measure is so very complex that few people could understand the equations. As a practical Electrical Engineer, I am used to building things that work. Most often it is not necessary to fully understand the detailed mathematics of the things that I build. Once you understand how things work that is sufficient to basically understand the object you are studying. The same is true of the universe.

It is important to note that although very complex wave equations are used by scientists to describe mathematically how things work, simple models and simple equations suffice to work as well. It is also important to understand that the universe that we see and measure is governed by things that we cannot readily see and measure. Therefore there is a set of simple laws upon which the universe operates which control how the universe that we see and measure works.

This book brings to light the Dot-wave theory which is a model of the basic structure of the physical universe. For this model, the entire physical universe is composed of a multiplicity of just three different kinds of dot-waves and three different types of energy. There are plus dot-waves and minus dot-waves. These are electrically charged dot-waves. There are also bi-polar dot-waves which are photonic dot-waves. From these three different types of dot-waves everything physical is made.

The dot-wave is a transformational device. It oscillates between light speed Co and light speed Cs.

The dot-waves take three different energy forms. We are all familiar with the photon which is a planar wave. The photon spins around a plane and at the same time moves forward at the speed of light Co. It is pure energy and has no mass until it can be converted into the two other forms of energy.

The Dot-waves can form planar patterns of energy. We can call this rotational energy or angular energy. The Earth moves around the sun with angular energy. Every split second in time the Earth has linear photonic energy but since it is acted upon by a force in a perpendicular direction it has resultant angular energy.

The third form of energy is spherical energy. The electron moving around a proton forms a spherical pattern. This caused the hydrogen atom to have additional mass when more linear photonic energy is added to the electron which causes its angular speed to increase. In addition, a spherical object that contracts and expands cyclically has spherical energy.

The dot-waves themselves possess spherical energy. They expand to a radius at the speed of light Co. Then they compress toward a pinpoint. At the pinpoint they move into the Cs dimension. There they expand to a larger radius and later compress back to a pinpoint. Then they reenter the light speed Co dimension.

If we removed all light speed Co/Cs and Cs/Co dot-waves from our universe, what would be left? The answer is that space would be left. What is space? What is the basis of our existence? What causes gravity? How was the physical universe produced? All these questions will be explained in the chapters to follow.

The Dot-wave theory models will help to explain gravity from a light speed Co/Cs perspective. It will give us the time of the Universe since the big bang. It will give us the state of the universe from before the big bang. The reader will learn from simple algebraic equations and Engineering models which are readily understood.

Most of the book is devoted to the original 1981- 1983 Dot-wave theory models which were explained from a purely light speed Co perspective. These models are shown in my "Doppler Space Time" © 2000 book.

Added to the original chapters is the understanding of the dual light speed Co/Cs dot-waves. It is the oscillating light speed Co/Cs dot-waves which causes the force of gravity to exist.

Finally after "The Dual Light-Speed Universe and the Dot-Wave Theory" was published in early 2018, the problem of Quantum Entanglement was studied. This caused a relook at my work from 1981-3 in which it was noted that the gravitational constant G and the permeability constant Uo had the same units. This meant that the ratio of the two constants of 18832.84 was an invariant constant of the universe. All the other constants are variable except for the light speed Co.

From this understanding, the higher light speed Cs was discovered where Cs= 18,833Co.

My work involved pencil and paper and a pocket calculator to solve the riddle of gravity. The scientists have gigantic computers and huge amounts of fellow scientists to advance my work to their level. They will be able to produce improved models. The scientists will improve things for sure but first they must learn more about the light speed Co/Cs universe and space itself.

This work is the result of 37 years of independent study. To me it was a great challenge to understand gravity, space and time. I now share my effort with those who seek to understand the universe from the creative mind of a practical nuts and bolts engineer.

Table of Contents

Introduction

I started looking for a solution to Gravity and Doppler Space Time in 1981 at the age of 42. The questions of the fundamental nature of physics came into my mind as I went to college at Polytechnic Institute of Brooklyn N.Y. from 1956 to 1966. When I graduated B.S.E.E. (summa cum laude) I went to work for various companies and retired from the Radar Research Dept. of Sperry Rand/Unisys in 1993.

In 1981 I started to question why the physics of the universe involves simple algebraic equations from the classical point of view and very complex equations from the modern physics point of view. Is there a fundamental structure of the universe of extremely simple nature upon which the complexities of what we see and measure exists? My study involved looking for the physics of the universe that we cannot see and measure upon which the physics of the universe of what we can see and measure exists.

To do that we must find the fundamental particle/wave of the universe upon which everything depends. The important question involves the fundamental structure of the universe. Quantum mechanics provides a mathematical solution which together with Einstein's relativity gives the physicist pieces of a general theory which do not perfectly match. The quantum physicists end up with various sub-particles which provide a basis for the particles and photons of the measurable universe. At CERN we look for more answers but the main problem is that most of the answers that are obtained are those that can be readily measured. This limits us to reactions between electrons and sub-particles with the electromagnetic fields we produce. In spite of that we can produce some experiments which indirectly enable us to understand some of the mysteries of the greater universe.

Now we are using a large machine to detect gravitational waves from rotating dual black holes. This gives us a clue as to the nature of gravity. The scientists are hoping to find that the gravitational radiation is due to a gravitational particle. The purpose of this book is to explain the light speed Co/Cs dot-wave which is the fundamental particle/wave of the universe. It is the source of the property of mass and the gravitational and electromagnetic fields.

One important question is whether the universe that we live within is a purely electrical universe, a mechanical universe, or a combination of both. The electrical equations and the mechanical equations tend to appear interrelated. Einstein's constant light speed space-time solution appears similar to what happens within an electrical field. Therefore it appears that the electric fields and the gravitational fields are related characteristics of Co/Cs dot-waves.

If this is so then the units of kilograms, coulombs, meters, and seconds can be replaced with only coulombs, meters, and seconds. Then when we look at the entire universe we find that it is composed of just two basic things called plus dot-waves and minus dot-waves. When the dot-waves are combined equally they form gravitational fields and uncharged particles. When they are unbalanced into plus and minus dot-waves, they form electromagnetic fields, charged particles, and sub-particles.

My work initially involved looking at the universe from a series of conversion charts in which the unit of kilograms was replaced by various combinations of coulombs, meters, and seconds. The charts were then studied for a few years. In addition numerical analysis was used to find the relationships between the various constants of the universe. The result of this work was to produce various sister solutions and some very basic equations of the universe.

It was my belief at that time that the relationships between the constants of the universe involved simple numbers such as 2, e, pi, 4, 4pi, 16 pi e. and 137.036. I spent many long nights trying to find combinations of numbers which would match these requirements. Therefore my initial work involved looking at conversion charts and numbers.

This effort caused me to believe that the entire universe was composed of only two things which I initially called dots. Years later I called them dot-waves. There were plus dots and minus dots. I formulated simple equations to find the charge of a dot and the mass of a dot. All along I was studying my conversion charts and my numerical relationships.

I would fill my mind with all my ideas. My mind has always worked best on new ideas as I slept. During the night my inner mind would solve the problems and I would awaken with the solutions. In the day time at work I would have many problems to solve. I would not find an answer right away. However the data went into my brain. I would go to sleep and during the night my high light speed Cs/Co inner brain would work on the problem and transfer the solution to my physical Co/Cs outer brain. I would awaken with the answers.

Sometimes during a test I would lose the simple light speed Co/Cs thinking process and my outer mind would be taken over by my high light speed Cs/Co inner mind. I would just write the answers down and they were always correct.

I just always accepted that it was a reaction between my normal mind and a higher dimension within me. All I knew for sure is that I observed different levels of my mind within me. There is my regular mind and there is a higher high speed mind that takes over when I am in an excited state. In addition this higher mind works when I sleep.

Perhaps I am unusual or perhaps what I experience is common to many people. In any event there is always a conflict between the various sections of my mind. The minute I have one possible answer, my mind swings to other possibilities. Every time I think I have the answers, my mind improves on them or destroys them. Thus I have a constant struggle between my everyday mind and my higher mind as I seek to find ultimate truth.

Over the years I continued to write books and sought the answers to my questions. The minute I would finish a book I would find flaws in it. My theory was not good enough. My understanding was not good enough. I was always looking for new and different possibilities.

This 80th year has given me new dreams and new ideas to study. Suddenly the dual light speed oscillating dot-wave became evident to me. It is the oscillation of the dot-wave between the light speed Co dimension and the light speed Cs dimension that produces the property of mass from an electrical charge. This was the missing link that I sought all these years.

Soon I will be gone from this Earth. I am finally starting to understand gravity and the dot-wave theory. My mind is starting to quiet down. My purpose is not to bring mankind an exact scientific understanding of the universe. My purpose is to bring to mankind an Engineering level understanding of the universe. From this Engineering level understanding of the universe, future scientists and mathematicians can provide more exact mathematical answers.

Since my understanding is at an Engineering level, this will enable the average scientifically inclined individual to understand the workings the universe. The mathematicians and physicists can move beyond this level of understanding to a higher scientific level. However the average person hopes to have a general understanding that is clear to him.

There are few people who could understand Einstein's complex space time equations. Likewise there are few people who could understand Quantum mechanical theory. As an Engineer these things never interested me as well. I believe that the fundamental workings of the universe can be adequately expressed by using simple algebraic equations.

As I learned about electricity at Brooklyn Technical High School, and later at Polytechnic Institute of Brooklyn, it appeared to me that the laws of electricity were sufficient to explain most things. As I looked at the moon I came to understand that the moon did not spin as it rotates around the Earth. It became obvious to me that the face of the moon was locked to the Earth by some sort of electrical type force. Thus the gravitational photonic field behaves much like an electrical motor.

By 1983 I had produced a version of the dot-wave theory with the basic equations of gravity in it. Over the years from 1983 to 1993 I worked on the theory part time. In 1993 I retired at age 55 during a company downsizing and was free to pursue my theory. By 2000 I self-published "Doppler Space Time" and went to work on my other books. I did little new work on the theory until late 2014 when I looked at it again. In 2015 I self-published "Gravity and the Dot-wave Theory" and for a while I was happy with the book.

However I learned that scientists measured gravitational waves coming from spinning sets of black holes. At the same time my theory did not answer some questions that bothered me. This caused me to relook at the theory again and a new understanding came to mind. In 2016 I self-published "Relativity and the Dot-Wave Theory" followed by "The Gravitational Wave and the Dot-wave Theory".

In the late winter of 2017 I started to study the possibility of some interactions between the light speed Co dimension and the light speed Cs dimension. I came to believe the possibility that light speed Cs had a speed of 1000 light years per second.

I then wrote a series of three books for the dual light speed universe. The physics book was "The Dual Light-Speed Universe and the Dot-Wave Theory. I completed these before February 2018. After wards I saw a TV program which described new measurements in Quantum Entanglement. I started to study the work of several scientists. Shortly thereafter it became self-evident to me that the ratio of the Gravitational Constant G to the Electrical permeability Constant Uo was the ratio of the light speed velocity Cs to the light speed velocity Co.

This necessitated the rewriting of the Dual Light-Speed Universe and the Dot-Wave Theory to include quantum entanglement.

My effort has always been a work in progress but in my 80th year the last three books was supposed to be my last attempt to understand gravity and the structure of the universe. It took me thirty seven years to reach this stage. I was hoping that my effort will enlighten the scientific community with new ideas and new thoughts. This will enable them to move upward even higher than I reached. I believe that I have broken new ground and it is left to others to move forward in the direction of higher knowledge and higher truth.

After the three books were completed at the end of winter in early 2018, I went to work on a "She Shed" for my daughter. The basic shed was installed by a carpenter and I finished insulating it and adding plywood inner walls. Now it has heat, AC, television, and furniture. I enjoy handyman work and it is a great addition to the property that I rent from my relatives.

After I finished the shed my high speed inner mind became active again. What did the ratio of Uo and G mean I wondered? I did not understand what it meant in my 1981-3 work. However my inner mind recognized the meaning of the ratio of Uo to G.

The book starts with Chapter One, "The Dimensions of the Universe" which discusses the three spatial dimensions of X, Y, and Z. In addition it discusses the two light speed dimensions of Co and Cs.

Chapter Two, "The Unseen Universe" presents an understanding of the universe such that the basis of the universe depends upon things that cannot be readily seen or measured.

Chapter Three, "The Hydrogen Atom" illustrates that the stability of the Hydrogen atom depends upon Einstein's mass to energy equations and Quantum mechanical wave equations. In addition the chapter explains the dual light speed analysis of the hydrogen atom.

Chapter Four, "The Three Energies of the physical Universe" looks at the three different types of energies that comprise the universe. This consists of linear, angular, and spherical energy. Both linear and angular energies are well understood. Spherical energy is the energy that is the basis of mass and gravity.

Chapter Five, "Conversion of Mass to Charge Velocity" shows the conversion of kilograms to coulomb meters per second. Here the Sister one and Sister two conversion possibilities are explained along with the conversion theory. The Sister one solution enables us to readily convert the gravitational constant in terms of the permeability constant. It also enables us to write the equations of the gravitational attraction between two hydrogen atoms. From this equation the age of the universe since the big bang was calculated to be 13.78 billion years.

The chapter uses the Eddington number of equivalent neutrons to calculate the mass of the visible universe and to total mass of the universe from the latest astronomical data. The mass of a dot-wave is calculated along with the amount of dot-waves within the electron, proton, and neutron.

Chapter Six, "The Dot-wave Theory" looks at a model universe and the variation of the mass of the light speed Co/Cs dimension of the universe with time. It also calculates the variation of the constants of the universe as the universe expands.

In addition, the normalized cycle time of the universe is calculated from the permeability constant Uo and the permittivity constant e_o.

Chapter 7, "Linear Doppler Space-time" is explained from the point of view of the Doppler Effect. The root mean square of the Doppler Equations is identical with Einstein's equations. The importance of Doppler Space Time is that an object in motion has a larger forward mass and a smaller rearward mass. This helps to explain the property of inertia. In addition the forward length of the object is shorter and the rearward length is larger.

Chapter 8, Angular Doppler Space-time" discusses the space time effects of revolving masses. It shows how the Doppler Effect builds up pressure in a rotating cylinder or wheel that causes it to fly apart.

Chapter 9, "Spherical Doppler Space-time" discusses how mass is produced by spherical type oscillations or equivalent angular oscillations with perpendicular components.

Chapter 10, "Space Travel" discusses the effects of Doppler distortions on the spacecraft and the human body. It discusses the proton thruster engine that takes the energy from the proton to propel a spaceship up to 20 percent of the speed of light

Chapter 11, "The Light Wave and the Photon" discusses a model of the photon. It discusses the variation of the energy of the photon due to gravitational fields and the relative motion the sun and the earth.

Chapter 12, "Black Holes and other Interesting Things" discusses black holes, high speed revolving stars and other interesting things.

Chapter 13, "The Multi-light-speed Universe" discusses the possibility of a series of universes of higher and higher light speed dimensions which reach out toward infinity.

Chapter 14, "The Gravitational Field" explains the details how the bipolar dot-waves produce the gravitational field.

Chapter 15, "The Electromagnetic field" explains how the plus and minus dot-waves produce the electric field and the magnetic field.

Chapter 16,"Conservation of Intelligence" explains how the intelligence within the Cs/Co dimension flows toward the pinpoint during the contracting cycle, and then flows into the expanding Co/Cs dimension during the expanding cycle. The intelligence is conserved and this will cause the man and life to repeat over and over again forever.

The Chapter discusses the dual light speed mind of man and the physics of the cosmic reincarnation process from a purely scientific viewpoint.

Short Biography of the Author contains salient details of the life and work of the Author.

Appendix: Provides a table of U.S. standard values and a table of calculated values.

Index of Selected Topics: Provides an index of important topics.

-------   ----2Ru

-------   -----Ru

-------   --Center

The picture shown is that of a homogeneous non-rotating universe. We live on a spherical plane at a distance Ru from an absolute center. The spin of the universe and non-linearity's will change the spherical plane into a non-linear ellipsoidal plane.

The outer radius of the universe in dark matter/ dark energy will always be a perfect sphere. It is the rotating center sphere that exhibits non-linearity of size and shape. In addition there will be both axial velocities and rotational velocities that produce many distortions.

Chapter 1: The Dimensions of the Universe

Section 1:0 Introduction

In classical physics we had three dimensions. These were the spatial dimensions of X, Y, and Z. Things were described in terms of distances and a time clock. Time was not a dimension.

Einstein came along over a hundred years ago and shocked the physics community with his time dimension. He gave us space time. Thus we lived in a four dimensional universe.

Over time Einstein's equations worked very well in describing the universe. Time clocks did slow relative to the Earth as they spun around in satellites.

Quantum theory originally had three spatial dimensions. However many quantum scientists now believe that the universe has four spatial dimensions.

String theory came along and the various mathematicians had ten or eleven dimensions. Space was filled with various small dimensions. We are basically a three spatial dimensional universe which is part of a higher dimensional universe. Thus the math permits many coexisting universe to exist within the same X, Y, and Z space.

Section 1-1 Einstein Space Time verses Quantum Theory

There is a conflict between Einstein's General Relativity and Quantum Theory. Quantum entanglement measurements indicate interactions at speeds greater than 10,000C. Einstein limits the speed of the Universe at the speed of light Co.

Super inflation at the big bang is the only plausible reason for the universe to appear at it is today. Thus the early universe expanded much faster than the speed of light Co. Thus it appears that Einstein's theory is an excellent approximation for understanding a universe in terms of the light speed Co and the time dimension.

Some quantum physicists believe that the time dimension does not exist. This would invalidate Einstein's relativity. In spite of this Einstein's work is quite excellent. It is an electrical theory approximation for the way the universe works.

Section 1-2: What is Time?

Classical physics described things in terms of X, Y, Z, and T. The time clock was very important. Somehow the universe came into existence at a particular time. We are born and our lives are measured by a time clock and calendar. So our brains accept that time exists alongside the spatial dimensions.

Einstein turned time into a dimension as he described the universe in terms of space time. Everything had a first cause. The universe was created by a first cause. The problem with that is what produced the first cause? Therefore the cause and effect problem only appears if time is an actual dimension. Once we eliminate time, there was no first cause.

Without time, the universe always existed. It was not created now will it end. The big bang was not a first cause of our universe. It was merely a point of inversion in which energy flowed into a pinpoint and produced the universe.

Without a first cause, energy always existed. We cannot ask how energy got here because energy is without beginning nor end.

The universe is an endless circle. It existed forever and will continue to exist forever. It merely changes from one particular configuration to another forever.

Section 1-3: Dimensions of the Dual Light Speed Universe

The dual light speed universe has five basic dimensions. There can be many more since many universes can coexist simultaneously. There can even be interactions between the various universes.

We have the X, Y, and Z spatial dimensions. We do not have the time dimension. We have the lower light speed C_o dimension and the higher light speed C_s dimension. According to the dual light speed analysis:

(1-1): C_o = 2.99792E8

(1-2) $C_s = U_o/G \, C_o$ = 18832.8

Where U_o = 1.25664E-6, and G= 6.67260E-11

(1-3) C_s = 5.64592E12

We live in a dual light speed universe. Our primary light speed is C_o and our secondary light speed is C_s. For our universe:

(1-4) Time = Distance / Light speed

Instead of the time dimension we have the light speed dimension. Thus Einstein's work could be expressed in terms of distance and light speed.

For the dual light speed universe the dot-waves oscillate between the light speed C_o universe and the light speed C_s universe. This causes effects similar to Einstein's work as dot-waves traveling more in the C_s dimension will appear to have moved through contracted space time. Their time clocks will be slower as well.

3

We then have to speak of the percentage of distance traveled at Co and the percentage of distance traveled at Cs.

The problem with this is that it is much easier to discuss things in terms of time. We can say that a proton is traveling at Co for a certain percentage of the time and at Cs for a certain percentage of the time.

To make matter more complex, we are dealing with probability. We are also dealing with transient dot-waves. The dot-waves oscillate from light speed Co to light speed Cs. There will be a transient period as they pass through the pinpoint between the Co and Cs dimensions.

The math of the dual light speed universe is more complicated than Einstein's Relativity and Quantum Theory.

This math is for future scientists and mathematicians to handle. All I can do is produce engineering type models to explain how the universe works.

Once future mathematicians and scientists understand how the universe works, they can produce the detailed scientific analysis of the universe.

The work presented herein uses simple algebraic equations and simple models to explain the universe. This will enable the scientifically inclined reader with modest math skills to fully understand the basics of the dual light speed universe.

Chapter 2: The Unseen Universe

The physics of the universe is based upon what we can see and measure. We build telescopes to look far back into the universe from the time it came into existence. We use microscopes to look at the smallest things. Our measuring instruments permit us to study electric and magnetic fields. The net result of our observations and measurements are a set of laws which describe how the universe operates from the level of what we can see and measure.

Einstein's relativity and Quantum Mechanics, provides us with theories and formulas which enable us to understand many things. Classical physics enabled us to advance from the Stone Age into the present. We went to the moon with a very simple computer and the use of slide rules. Modern thoughts gave us fantastic computers and modern medicine.

All the equations and theories are bits and pieces of a general theory which cannot be made to match all the individual theories and equations. Quantum Mechanics gives us a great theory for the working of particles and sub-particles. The Quantum scientists attempt to explain things in terms of various sub-particles. They seek to find a sub-particle which will explain gravity. Something is missing and they will never find the missing link using single light speed theory. However slowly they are advancing as they study quantum entanglement experiments.

They can build the greatest machines such as at CERN and produce the greatest electro-magnetic fields. In so doing they can produce matter and antimatter, yet they can never produce a sub-particle that is responsible for gravity. Something is clearly wrong with the various theories.

When we look at Einstein's equation for the conversion of mass to energy:

(2-1) $E = MC^2$

In equation (2-1), Einstein and others found that a material could lose mass with the result that a lot of energy would be created. This simple relationship illustrates that the universe operates upon some very simple algebraic laws. This equation is a fundamental law of the universe.

The law simply states that mass in terms of spherical type energy converts into photonic energy which is linear type energy. The C squared term illustrates that oscillatory spherical type energy when released from an object sends out this energy at a very high speed. This equation is the basic equation of the universe.

The Dot-wave theory specifies that everything in the physical universe is built from a multiplicity of only three things and three different energy forms. There are plus dot-waves, minus dot-waves, and bi-polar dot-waves. In addition each thing can occur within three different forms of energy. There is linear energy, planar energy, and spherical energy. To make matters more complex we can get things composed of multiple combinations of the three things and three forms of energy.

The dot-wave theory helps us to understand the physical universe that we see and measure. Einstein's theories and Quantum mechanics provide us with good mathematical analysis of the workings of the universe.

The problem that we are faced with is what caused the present universe to come into being and what holds it together? No matter how hard we try to understand these things we fail to convince ourselves of the truth of our findings. Once we come to understand that time is only an imaginary construction of our physical reality, we no longer have cause and effect to concern ourselves with. The universe has neither beginning nor end. It merely moves from one configuration to another. Thus the present universe is a primary Co/Cs universe.

The universe prior to the big bang inversion was a Cs/Co universe. This gave it a much larger maximum radius before the compression. It is possible that there was a spectrum of universes leading toward light speed infinity for Cs*. That is for philosophers to ponder. I believe that is true but the present book illustrates the physics of the dual light speed universe.

Our dot-waves of light speeds Co/Cs are the product of dot-waves of light speed Cs/Co. The universe that we can directly see and measure is the product of the universe that we cannot directly see and measure. However by observing the Co/Cs universe we obtain indirect measurements of the light speed Cs/Co universe. In this book we can look at the Co and Cs dimensions as universes or dimensions.

The important thing is that the two dimensions are interrelated with each other. A Co/Cs dot wave switches every split second to a Cs/Co dot-wave. The large universe did a large inversion at big bang whereas the individual dot-waves continuously invert every split second in the nanosecond range.

The physical Co/Cs universe is the product of the higher light speed Cs/Co universe. We then have a dual light speed universe comprising of both lower and higher light speed energy.

My work on the original dot-wave theory in 1981-3 looked at the dot-waves as single light speed entities. This provided us with many interesting properties of the physical universe. Now we must look at the more general case of a dual light speed dot-wave to help us understand the universe better.

In the "Conversion of Mass to Charge Velocity" chapter, the most likely Sister One solution shows us that mass in kilograms equals coulomb meters per second. This causes both the gravitational constant G and the permeability constant Uo to have the identical units of Met^2/Coulomb seconds. This conversion along with many other sister solutions was studied in my work of 1981-3.

As shown in equation 1-2, the ratio Uo/ G is:

(2-2): $Uo/G = 18832.8$

Where Uo = 1.25664E-6 and G = 6.67260E-11.

For the constant light speed Co solution, all the other constants of the universe vary as the universe expands from a minimum radius toward a maximum radius. The speed of light Co of 2.99792E8 meters per second is a constant as meters and seconds both track each other.

That is what I believed in 1981-3. It seemed to me then that time was an independent variable. Somehow the universe depended upon time clock. Einstein believed that as well as he produced his space-time theory. Yet today it is self-evident to me that this is incorrect thinking.

My prior thinking specified that the units of time and seconds tracked each other. This gave us a constant light speed solution for the Co universe. However once we eliminate time as a variable, we get:

(2-3): Distance = meters

(2-4): Time = Distance/ Co = meters/Co

(2-5) Distance/Time = Co

8

The net result is that once we eliminate the units of time in the ratio of distance to time we get light speed Co. To say that the speed of light Co is a constant is an absolute certainty once we eliminate time.

Any variation in the universe can only occur when the dot-waves switch from light speed Co to light speed Cs. To say that time slows or the ruler expands is likewise meaningless. All we can say is that a dot-wave or large group of dot-waves moves from light speed Co to light speed Cs over smaller or larger distances. Thus everything in the Co/Cs universe continually moves between the Co and Cs dimensions.

The question of the meaning of the ratio of Uo/G was unanswered by me in 1981-3. As will be explained later, the speed of light is invariant and the ratio of the permeability to the gravitational constants is also invariant as the universe expands and contracts.

Scientific analysis of the quantum entanglement problem has shown interactions to occur at a speed greater than 10,000 Co. Therefore recently it became self-evident to me that Equation (1-2) represents the ratio of the higher light speed Cs to the lower light speed Co. Therefore:

(2-6)): $Cs/Co = 18832.8$

(2-7): $Cs = 5.64592E12$ meters/second

The dot-wave is a transformational entity. It can exist in the light speed Co dimension with a mass M_D and energy E_D. Likewise by using Equation 2-1; we find that the same dot-wave can exist in the light speed Cs dimension with the same amount of energy but with a mass of:

(2-8): $M_{DCs} = 2.81949E-9 \; M_{DCo}$

9

In Equation 2-8 we see that the dot mass when it is in the Cs dimension is 2.81949E-9 times as low as when it is in the Co dimension.

Everything in the universe exists within the low light speed Co universe and the high light speed Cs universe. Every dot-wave, every atom, and every higher structure has two different but interconnected existences. We live in the light speed Co dimension but we also live in the light speed Cs dimension. That will give philosophers more ideas to ponder.

Chapter 3: The Hydrogen Atom

Section 3-0: Introduction

The purpose of this Chapter is to show that the light speed Co/Cs universe that we see and measure is driven by Einstein's Space-time type energy equations and other basic equations. The Hydrogen atom is an example of this.

When we look at our universe we find that the basic structure of our universe is the proton, the electron, and the neutron. Quantum theory has explained the sub-particles quite well so that we understand the basic structure of the proton, neutron, and various neutrinos. Yet all these things in turn are composed of the dual light speed dot-waves.

The answers that we are looking for are what produces the property of mass and what produces the force of attraction between objects. The electric and magnetic fields are quite well understood but the gravitational field is not. Therefore the study of the hydrogen atom will help us to begin to understand these things.

The dot-wave theory specifies that everything is made up of dot-waves. Later we will calculate how many dot-waves are in an electron, proton, neutron, and photon. One thing that we know about the hydrogen atom is that it is stable. We have to add 13.8 electron volts to release the electron from the proton. Therefore the electron is bound to the proton within the hydrogen atom. We begin our study with the Bohr orbit.

SECTION 3-1: THE BOHR ORBIT

Let us now look at the velocity of the electron in the first Bohr orbit state. Bohr did an excellent job of bringing an understanding of the hydrogen atom to man. He compared the light spectra and wave theory equations to relate the

11

electric field and the centripetal force. He derived a set of equations, which fit into each other. He did not explain the basic reason for the binding energy of the hydrogen atom. He used the measured value of 13.58 electron volts to verify his work. No doubt his work was quite excellent. Today the modern binding energy on the Internet is listed as 13.60 electron volts.

Let us now look at the Bohr orbit from an Einsteinian viewpoint. Bohr deduced that the first orbit would operate at a speed of C/137.036 and a radius R_B at 0.529177E-10. (This is a modern number from U.S. Government data). He showed that there was a constant relationship between the square of the velocity of the orbit and the radius such that:

(3-1): $M_E(V_B^2)R_B$ = Constant

For an orbit of the series R_B, $4R_B$, $9R_B$... the velocities would be V_B, $V_B/2$, $V_B/3$.... The proton electric field attracts the electron and the electron heads toward the proton. Bohr had no explanation as to why the electron would not be captured by the proton. Yet his basic answers were great. The problem is that wave analysis is excellent at representing the mathematical properties of particles and sub-particles but it fails to physically explain the reason for these relationships.

As the electron moves toward the proton, its Einsteinian mass increases, which causes additional gravitational repulsive forces in the direction away from the proton. This prevents the electron from flowing into the proton.

Let us now look a little more carefully at the electron in the Bohr orbit at V= C/137.036 to see the relationship between the binding energy and Einstein's mass increase for the electron in the Bohr orbit. Let us use Einstein's formula:

12

(3-2): $M = M_o/[1-(V/C)^2]^{0.5}$

When V/C equals 1/137.036 we get

(3-3): $M = M_o /0.99997337 = 1.000026627 M_o$

The differential mass increase is:

(3-4): Differential mass increase = 26.6274 ppm

In Equation 3-4 we find that the electron has increased in mass by 66.6274 parts per million. The electron has a mass/energy of:

(3-5): Mass/Energy of Electron = 0.510999MEV

(3-6): Differential mass/energy in Bohr Orbit = 13.606 EV

In equation 13-6 we find that the electron in the lowest state of the Bohr orbit has an Einstein mass increase equal to the Binding energy of the electron. Thus:

(3-7) Binding Energy = Einsteinian mass increase

In order to release the electron, you would have to add a photon of 13.6 electron volts to stop the stop the electron from orbiting the proton.

What this tells us is that the Hydrogen atom is working on both Einsteinian energy relationships and Bohr wave analysis.

As the electron revolves around the proton it produces a spherical wave pattern which is similar to regular spherical waves. As this occurs we get a component of pure spherical energy. Thus a part of linear and angular energy converts into spherical mass within the hydrogen atom. Einstein's mass to energy formula works quite well for the Bohr orbit. Thus the hydrogen atom is a mini atomic energy converter. It illustrates the correctness of

13

Einstein's mass to energy and increased mass with velocity formulas.

It is necessary to hit the electron with a photon to nullify its momentum and slow it up as it moves to a higher orbit. At each step of the way the atom experiences conversions from spherical energy to linear and angular energy where angular energy is either angular or rotational energy. They are actually different forms of the same thing. Rotational energy can be considered a huge amount of little particles spinning in an orbit. Of course each form of the three energies can be subdivided by mathematicians into various special cases.

Once 13.60 EV is added to the electron it will convert its spherical image and associated gravitational mass into pure linear energy. Then it will be free. If we add even more energy to the electron, it will not only be free but will have a new velocity in a linear direction.

The importance of this chapter is to show that it is Einstein's equations that govern the most important structures of the Universe. The energy relationship of the electron within the hydrogen atom is due to the mass increase as defined by Einstein.

Bohr showed that the stability points in the hydrogen spectrum were based upon wave analysis. The net result is that the hydrogen atom works on both Einstein's principles and Quantum mechanical analysis. Einstein provided the energy and Bohr and company provided the stability points.

Within the hydrogen atom, each Bohr orbit has different values of spherical, angular, and linear energy. The mechanism is basically the mass to energy Einsteinian relationships but the actual states of the atom are wave analysis. This again ties Einstein's theory to Quantum mechanics.

Einstein and Bohr could not perfectly agree with the basic laws of the universe. Einstein looked deep into the mathematical universe while Bohr looked at the universe that is seen and measured. It appears that both were right in their work but both were working with two different models of the universe. Therefore Einstein's work and Quantum mechanics are not at odds with each other. They merely look at the universe from two different models.

Einstein's work was excellent from a mathematical analysis. However it did not grow much beyond Einstein. Quantum theory continued to grow over the years. However neither provided an adequate understanding of gravity and the basic structure of the universe.

The important thing to understand from this analysis is that the universe operates upon energy conversions. The energy conversions are the driving force and the wave analysis follows the energy conversions. There are two different ways of looking at the results but the laws of physics are driven by energy conversions. The same is true of the dual light speed dot-wave.

Section 3-2 The Bohr Orbit from Dual Light-Speed analysis.

From the above analysis we see that the hydrogen atom operates from both Einsteinian energy conversions and Bohr wave analysis.

From atomic experiments, the proton has a radius of:

(3-8) Rp = 0.8E-15

The measured radius of the proton is only 0.8E-15 meters in the Co dimension. The radius of the proton as it oscillates into the Cs dimension is:

(3-9) Rp* = Uo Rp /G = 18833 Rp

(3-10) Rp* = 1.507E-11 meters

The Bohr orbit radius is:

(3-11) R_{Bohr} = 5.29177E-11

From Equations 3-10 and 3-11 we find that the proton's light speed Cs radius is approximately 28.5 percent of the Bohr orbit. In addition the electron is larger than the proton and oscillates between Co and Cs as well. The net result is that the Co/Cs oscillations of the proton and electron determine the waves that Bohr recognized in his studies.

It has always been a puzzle why a tiny proton is circled by a tiny electron at such a relatively huge distance. It would appear that so much empty space exists. However for the dual light speed solution, the proton in the Cs dimension is 28.5 percent as large as the Bohr orbit and the electron makes up the remaining size difference.

Therefore the size of the Bohr orbit is governed by the light speed Cs dimension and not the light speed Co dimension. If we tried to push the electron into the proton the oscillating light speed Cs fields would repel each other. It takes a spherical compression to produce the neutron and a neutron alone would self-destruct.

Section 3-3: The Fine Constant Calculation

The fine constant can be calculated from numerical analysis by using a mathematical model. Let us assume that the electron rotates around the proton in 274 half-waves in one plane and 274 half-waves in a perpendicular direction. The inverse fine constant would be:

(3-12) (Fine Constant)$^{-1}$ = 137/Cosine (360/274)

(3-13) (FC)$^{-1}$ 137/Cos1.31386° = 137/0.999737

(3-14) $(FC)^{-1} = 137.036$

From this numerical analysis it appears that the electron moves around the proton in 274 half waves in one direction and 274 half waves in a perpendicular direction.

The net result is a shift of 1.31386 degrees each time. This causes a wave pattern with 137 parts to have a wavelength just a little bit larger thereby producing the 137.036 factor.

Chapter 4: The Three Energies of the Physical Universe

The universe has three different types of energy. The first energy we are familiar with is for an object that moves in a straight line with a velocity V. Such an object has linear energy or linear momentum. The object will continue to travel in a straight line unless acted upon by a force in a perpendicular direction.

The Earth appears to travel in a straight line every split second but over time the perpendicular gravitational attraction of the sun causes it to move in an orbit around the sun.

The second energy we are familiar with is angular energy. A planet moving in a plane has angular energy. If a force acts perpendicular to the plane of the orbit, it will still have angular energy but the orbit will move in a more complex pattern around the center. The result is that a spherical angular pattern can be formed. However this is still a form of angular energy. A planet revolving on its axis is another form of angular energy. Every atom within the planet can be considered as orbiting around the center line.

The last form of energy is spherical energy. This energy takes the form of an expanding or contracting spherical surface. The spherical energy flows outward and or inward from the center. Often spherical energy forms take the shape of a spherical oscillation.

The mass/energy of a dot-wave is caused by a spherical oscillation. In general a neutron contains huge amounts of dot-waves so the radius of the neutron at light speed C_o is very small. When the oscillation passes through a pinpoint and moves into the C_s dimension, the radius of the neutron is much larger. However it is still near the size of the Bohr radius.

Spherical energy can also have a rotational component and this allows an object to have both spherical and angular components. In addition we can have an object moving in space in a straight line while rotating. Then all three forms of energy are present.

An electron that revolves around a proton in one plane and then the plane orbits in an orthogonal plane has a form of spherical energy. Thus there is a spherical energy component within the angular energy field.

The property of mass requires a spherical oscillation or a complex angular rotation around a point. The point may be stationary or moving as long as the linear speed is less than the speed of light Co.

A photon contains both linear and angular energy. It does not have spherical or equivalent energy and thus it is massless and not gravitational. It can travel at the speed of light but when it is absorbed by an electron; its energy tends to convert into linear, angular, and spherical energy. Therefore the physical universe constantly converts all three forms of energy into each other.

The electric field is a form of spherical energy in which photonic energy can change into positive and negative fields. If we move a spherical positive or negative field, we get a linear and an angular component that is the magnetic field.

Chapter 5- Conversion of Mass to Charge Velocity

Section 5-1: Introduction

In this chapter simple classical type equations for the Dot-wave theory will be used to find the conversion from mass to charge velocity. The universe is an electrical-mechanical universe and the units of kilograms can be described electrically using the correct space time conversion factors.

At present the universe is specified in terms of kilograms, coulombs, meters, and seconds. There are many different names for things such as Amperes but Amperes are really coulombs per second. In the usual mechanical/electrical physics books, four units are required. Once we convert mass in terms of coulombs, meters, and seconds then only three units are required.

The mass to charge conversion is limited to the constraints of the universe that we live in. If the universe is expanding with increasing ruler and time clock while at basically constant light speed, then we have fewer conversion choices.

Section 5-2: Conversion Theory

Most of the equations studied were a set of normalized equations for a model universe that expands at a basically constant light speed and with an expanding ruler and an expanding time clock.

Some of the equations studied are standard physics equations, whereas others are for the normalized model universe. These equations serve to reduce the equations of physics into three variables of kilograms, meters, and seconds; or coulombs, meters, and seconds. Using these equations, it will no longer be necessary to use combined electrical and mechanical units. Although this is important to study to understand the basis of the universe, the mechanical equations are much easier to use for ordinary mechanical problems.

In order to eliminate either kilograms or coulombs, it is necessary to produce a table of conversion. We can replace kilograms with some function of coulombs, meters, and seconds. Thus we can define mass in terms of charge and some power of light speed. Charts can then be produced using many combinations that match the equations in the dot-wave theory as far as units are concerned.

The process for the production of the charts is readily accomplished. The comparison study of the charts takes many years to accomplish. Each chart will produce equations for the universe that may or may not represent reality. Fortunately the charts tend to be dual or sister solutions so things learned from an incorrect chart still applies to our universe. Thus all the solutions can be considered Sister Solutions. However, only one solution is the actual solution and this is called the Sister 1 solution.

As a Sister 1 solution we could say that mass has the units of coulomb meters/second and that energy has the units of coulomb meters^3/seconds^3. This solution meets the Dot-wave theory criteria of basically constant light speed since as meters increase and seconds increase, the light speed remains constant.

The mass then only depends upon the total coulombs within particles, which decrease as dot-waves radiate from matter and the universe expands. This solution says that mass has the same units as a moving charge. It also says that energy is charge oscillating in a volume.

For a Sister 2 solution, a dual solution could be looked at. Mass would be coulomb seconds/ meter and energy would be coulomb meters/ second. In this solution charge would be a property of a moving mass.

In general for a Dot-wave theory solution, mass must equal charge times any power of light speed. Therefore,

(5-1): $M_D = Q_D C^n$

Where C^n could be any positive or negative power of the light speed. The corresponding dot-wave energy is:

(5-2): $E_D = Q_D C^{n+2}$

In equation 5-2 we see that once the form of the dot-wave mass is chosen, the energy is C^2 higher since $E=MC^2$.

We can now produce a chart of the Sister 1 solution to see the various relationships. In general the Sister 1 solution states that:

(5-3): $M = Q C$

Equation 5-3 states that the unit of mass is charge times the speed of light. This basically states that a moving charge causes the property of mass. However the charge does not move physically in spatial dimensions. It moves from the universe of light speed Co to the universe of light speed Cs.

In each light speed dimension, the dot-wave expands and contracts to a pinpoint where it moves into the adjacent dimension. The mass/energy is an expanding and contracting bubble for each dimension. This is the basic value of the mass. When we add linear photonic energy to the mass we increase the linear speed and the equivalent mass. However the basic mass/energy of each dot-wave within a mass is unchanged. The total mass/energy of an object increases as more dot-waves are added to the object.

When we move a charge from light speed Co to light speed Cs, the energy will stay constant and the charge will stay constant but the mass will change from a high mass at Co to a low mass at Cs. This is on the mass verses light speed operating line.

The dual solution or Sister 2 solution is:

(5-4): Q = MC

Equation 5-4 states that the unit of charge is mass multiplied by the speed of light. This basically states that a moving mass causes the property of charge.

For this sister solution the property of charge is caused by a mass oscillating between the light speed Co universe and the light speed Cs universe.

In this solution, the dot-mass would remain constant as the dot-wave oscillated between the Co and Cs universe. At the same time the dot-wave charge would change from a low value in the Co dimension to a high value in the Cs dimension.

I have investigated many other solutions over the years but the simplicity of equations 5-3 and 5-4 causes either one to be considered a most likely solution. However, the Sister 1 equations will be shown to match the physical universe more readily.

Charts of other solutions are easily prepared. However, once higher powers or square roots or cubes of the light speed occur, they lack the simplicity of the Sister 1 solution or the dual Sister 2 solution.

The important thing in the method shown is that the equations presented so far are unit's equations in which only the three ingredients of kilograms, meters, and

23

seconds, or coulombs, meters and seconds are used. Since the mechanical world interlocks with the electrical world at energy and force equations, it was never necessary to have so many different units.

At most we only needed kilograms, coulombs, meters, and seconds. By specifying that the universe is either completely electrical or completely mechanical, then only three units are needed.

Let us now look at a table of the most likely solution for the relationship between mass and charge.

The following table of units is made for the most likely Sister 1 solution with Mass = Charge Velocity.

Table 5-1: Conversion of Mass to Charge

Quantity	Sister 1
Mass (M)	Cou Met/Sec
Charge (Q)	Coulombs
Energy (E)	Cou Met^3/ Sec ^{3}
Coulomb Const.	Met^4/ Cou Sec^3
Force (F)	Cou Met^2 / Sec^3
Momentum (MV)	Cou Met^2 / Sec^2
Plank's Const. (h)	Cou Met^3 / Sec^2
Permeability (U_O)	Met^2 / Cou Sec
Permittivity (e_o)	Cou Sec^3 / Met^4
Voltage (V)	Met^3 / Sec^3
Current (I)	Cou / Sec
Impedance (Z)	Met^3 / Cou Sec^2
Grav. Const. (G)	Met^2 / Cou Sec
Power (P)	Cou Met^3 / Sec
Flux Density (B)	Met / Sec^2
Inductance (L)	Met^3 / Cou Sec
Charge/Mass	Sec/ Met
Capacitance ($\underline{C}$)	Cou Sec^3/ Met^3

In the table the various quantities have been shown in the Meters Coulombs Seconds (MCS) system for the most likely Sister 1 solution.

Section 5-3: General Mass to Charge Conversions

In Section 5-2, the Dot-wave theory method of matching mass to charge was formulated. The basis of the method was that charge and mass both decreased with time or distance as the universe radiated and expanded. The relationship between mass and charge then became ratios of the speed of light to various positive and negative powers.

A more complete solution of various possibilities would be that mass also could vary with charge or current and a power of the speed of light. Thus more possibilities exist beyond the Dot-wave theory although they become less probable. Let us look at the simplest general possibilities. The following are possible:

(5-5): Mass = Charge (M=Q)

(5-6): Mass = Current = Charge per second (M=I)

(5-7): Mass = Charge per meter (MC=I)

(5-8): Mass = Current x Velocity (M=IC)

(5-9): Mass = Charge /Velocity (MC=Q)

(5-10): Mass = Charge x Velocity (M=QC)

We can now make charts of all these conversions by using the following standard equations:

(5-11): Force = KQQ/R^2

(5-12): h = Energy x Time

(5-13): GMM = KQQ

(5-14): V = KQ/R

(5-15): $Uoe_o = 1/ C^2$

26

(5-16): B = Uo I/ R

Equations 5-11 through 5-16 are standard physics equations and enable us to produce charts of the various relationships in terms of meters, coulombs, and seconds for the MCS system. We can also chart the modified GG/MKS system of meters, kilograms, and seconds.

The solution for mass = charge, (Equation 5-05), tells us that a moving mass produces currents, and positive and negative magnetic fields from positive and negative dot-waves. Since mass is loaded with dot-waves, one characteristic of mass is identical to charge itself. However, when we look inside the proton from a dot-wave perspective, we see dot-waves in constant motion. We see current flows and they look like current gyroscopes. Thus a primary characteristic of mass is not charge but it is related to the motion of charge. Although (mass = charge) falls within the Dot-wave theory since both mass and charge decrease with an increasing universe, it is not a real possibility.

The solution mass = current, (Equation 5-06) is a characteristic of mass. If we look inside the proton, we do find currents within it. Yet, what give them the gyroscopic nature are not currents themselves but currents flowing in a circular path. This solution is not a Dot-wave theory possibility since mass and charge do not track each other as the universe expands.

The next possibility for the primary electrical characteristic of mass is that momentum is current, (Equation 5-07). Moving masses do have moving dot-waves and currents exist. That mass is a coulomb per meter can be considered a characteristic of mass. However, we are looking for a gyroscopic effect for the electrical equivalent of mass. This solution is not a Dot-wave theory solution but it is interesting to study and compare to the other solutions.

27

The next possibility is that mass is current times velocity, (Equation 5-08). In this case, as the universe expands the mass drops and the charge remain constant. This solution has all independent units. No basic relationship between the gravitational constant and the magnetic field is self-evident. It does provide an alternate solution; however it is not part of the Dot-wave theory.

The next possibility is that mass equals charge over velocity, (Equation 5-09), or that a moving mass creates charge as a dual Sister 2 solution. It provides a good electrical to mechanical analogy with mass and capacitance being equal. It remains the secondary dual solution for comparison study and analogy since the differential equations of capacitive circuits are of identical form as those equations for mass.

Let us now look at the last solution, (Equation 5-10) which is the Sister 1 solution. In this solution a moving charge produces mass. Thus:

(5-17): Charge x Velocity = Mass

Equation 5-17 is for charge momentum. It is the electrical dual of a physical momentum. Thus:

(5-18): Mass Momentum = Mass x Velocity

(5-19): Charge Momentum = Charge x Velocity

We see in equations 5-18 & 19 that both mass and charge momentums are similar quantities. Breaking down equation 5-19 into parts, we get:

(5-20): [Charge/second] x Meters = I x R = Mass

In equation 5-20 we see that we have a current (I) operating at a radius R that gives us a current torque or charge momentum. This is a strong possibility for mass since we can see within the proton, currents flowing in circular paths and oscillating from inner radius to outer radius as well. We also see a root mean square current at a particular radius. Thus the current gyroscopic action within the proton adds credibility to equation 5-20 as being a primary conversion equation.

Let us now produce a chart of the various relationships for this Sister 1 solution for the mechanical (GG-MKS system). In this system, the MKS system has been modified to show charge (Q) in terms of mechanical units where charge is kilogram seconds per meter.

TABLE 5-2: Charge to Mass Conversion (Q=M/C)

Quantity	GG-MKS-System
Mass (M)	kg
Charge (Q)	kg sec/ met
Charge/Mass	sec/met
Velocity	met/sec
Acceleration	met/sec^2
Energy (E)	kg met^2/sec
Force (F)	kg met/ sec^2
Momentum	kg met/sec
Plank's (h)	kg met^2/sec
Coulomb (K)	met^5/kg sec^4
Permittivity	kg sec^4 / met^5
Permeability	met^3/ kg sec^2
Grav. Const.	met^3/kg sec^2
Voltage	met^3 / sec^3
Current	kg/met
Impedance	met^4/ kg sec^3
Inductance	met^4 / kg sec^2
Capacitance	kg sec^4/ met^4
Flux Density	met/ sec^2

In Table 5-2 the gravitational constant has the same units as electrical permeability. This shows that gravity is similar to a magnetic force. In addition flux density (B) has the same units as acceleration. This shows that an accelerating space-time electromagnetic field produces magnetic flux. Since voltage is cubic velocity, this shows that the moving universe generates voltage. Thus the motion of the electromagnetic field produces currents and voltages and magnetic flux as it expands.

The relationship between the gravitational constant and the electrical permeability using a best numeric fit is:

(5-21): $G = 16 \text{ pi e Uo } / (137.036)^3 = 6.67223E\text{-}11$

Equation 5-21 shows the exact relationship between the gravitational constant and the electrical permeability constant. It shows that the gravitational field is purely magnetic in operation and that the conversion chart in Table 5-2 provides us with a very important conversion from mass to charge.

In the Sister 1 (MCS) solution mass per charge as per Equation 5-17, is a constant as the universe stretches out with both time and distance expanding. Energy is volumetric in both time and distance and drops as coulombs drop. Thus:

(5-22): $MV^2 = Q V^3$

In Equation 5-22, energy has the units of charge times velocity cubed. Likewise it is charge within the confines of cubic meters over cubic seconds. Thus energy is a volumetric space-time entity. It is charge oscillating in three dimensional distance (meters) volume over three dimensional time (seconds) volume.

From a dimensional perspective this appears to indicate that energy involves a vibration in the three spatial dimensions and the three time dimensions.

Now my words written in my "Doppler Space Time" © 2000 book have new meaning as the vibration is not within our universe alone but between the low light speed Co visible dimension and the higher light speed Cs invisible dimension.

Since there is no time dimension, we only have the X, Y, and Z dimensions and the light speed dimensions. When we move from Co to Cs and back again, each X, Y, Z dimension has a corresponding time component in terms of distance and light speed. Thus the original statement in the year 2000 is a simplified way of expressing the relationship.

This indicates that energy is charge times the speed of light cubed. Thus:

(5-23): Energy $= QC^3$

In equation 5-23 we specify that energy is the charge Q times the speed of light cubed. This means that when the dot-wave moves from the Co universe to the Cs universe, the charge of the dot-wave drops as the cube of the light speed. Thus:

(5-24): $Q_{DS} = Q_{DO} (Co/Cs)^3$

The mass of the dot-wave drops by the square of the light speed. Thus:

(5-25): $M_{DS} = M_{DO} (Co/Cs)^2$

The ratio of charge to mass of a dot-wave decreases as the dot-wave moves from the Co dimension to the Cs dimension. Thus:

(5-26): Charge to mass ratio from Co to Cs = Co/Cs

Now let us look at momentum

(5-27): $MV = Q V^2$

In Equation 5-27, for momentum, coulomb meters squared per seconds squared would be charge oscillating in two dimensional distance (meters) area over two dimensional time (second's) areas.

For the dual light speed analysis:

(5-28) Momentum = QC^2

As we move from the Co dimension to the Cs dimension, the charge drops by the cube of the ratio and the light speed increases by the square of the ratio. Thus the momentum drops by the ratio.

(5-29) Momentum ratio from Co to Cs = Co/Cs

Thus all the various units mean something with respect to the light speed dimensions which will be studied in Section 5-4.

It appears that momentum involves two dimensions X and Y. Energy involves all three dimensions and is volumetric while momentum is planer. This means that momentum is caused by photons moving in a planer fashion. In general photons only experience mass and energy when they interact with particles or strong gravitational fields. It has always been taught that photons appear massless but have energy. Evidently you have to occupy all three time dimensions in order to have mass. Photons only have momentum but no mass. They acquire mass when they attach to a mass and slow.

The Sister 2 solution has the simplest units. However, the more complex units of the Sister 1 solution provide us with a greater understanding of space and time

The fact that energy has the cubic units for the Sister 1 solution makes that solution more important. The Sister 2 solution is a great duality and since $M_NC=piQ$ approximately, it appeared most important originally. The Sister 1 solution matches at the gravitational constant / electrical permeability constant. However, that is what was sought since this is where the mechanical world matches the electrical world. Of course we must find what mass matches what charge.

The sister 1 solution enables us to see that mass is the spherical Co/Cs oscillation within the proton and electron. In the photon, you may get circular action perpendicular to the axis of motion but no circular action in the front to back region. Thus the photon will spiral and move forward. It will have energy in the front and rear as it oscillates between the Co universe and the Cs universe.

The five solutions investigated all are part of the general conversion equation from mass to charge. Thus:

(5-30): $GMM/R^2 = Uo\ QV\ QV/\ R^2$

Equation 5-30 is a general unit's equation of standard physics that relates the gravitational force to the electrical magnetic attraction. The units are standard physics. The five most probable solutions to this equation are:

(5-31): $G= Uo\ (M =QC)$

(5-32): $G= UoR\ (M = QC/\ R)$

(5-33): $G= UoC^2\ (M = Q)$

(5-34): $G= UoC^4\ (MC = Q)$

(5-35): $G = U_o R^2 C^2$ ($MC = QC/R$)

Equation 5-31 states that the units of G and Uo are identical. Both G and Uo can vary but they will track each other in both light speed dimensions.

Although it has been specified that there are five dimensions, each light speed dimension will produce the equivalent of three different times with respect to X, Y, and Z. Then we could say that there are three spatial dimensions and three equivalent time dimensions for the Co dimension and three equivalent time dimensions for the Cs dimension. That would change the number of dimensions to nine. This may be significant to the mathematicians and scientists but it is not part of my engineering models.

Equation 5-32 has G and Uo related by the distance R. This is a very unlikely solution. Equation 5-33 has the units of G and Uo related by the square of the light speed. Since the units of mass equals charge, this solution does not permit there to be any differences between the electric field and the gravitational field. It is not a feasible possibility.

Equation 5-34 with units MC = Q, is an interesting possibility but after much study, it is hard to relate it to our universe. The fact that the ratio of G to Uo is the fourth power of light speed makes such a universe unfeasible. Equation 5-35 is much too complicated to be a solution.

The net result is that after years of study of the various possibilities for the relationship between the electrical world and the mechanical world, only Equation 5-31 makes sense. This is called the Sister 1 solution.

Therefore for the rest of the book the Sister 1 solution will be used as the primary solution which relates charge to mass and the constants of the universe. We can then calculate the constants of the universe in terms of each other. Thus repeating equation 5-21:

(5-36): $G = 16$ pi e Uo / $(137.036)^3$ met^2/cou sec

Equation 5-36 is the best fit for the numerical conversion from the gravitational constant to the electrical permeability constant. This calculates to be:

(5-37): G= 6.67223E-11

In standard physics, the relationship between Zo and h and Q is:

(5-38): Zo = 2h/ 137.036 Q^2 = 376.729 met^3/cou sec^2

From the fine constant relationship in standard physics, we get:

(5-39): hC/ [2 pi K Q Q] = 137.036

As explained in Section 3-3, the 137.036 comes from the reciprocal of the cosine of 360 degrees over 274 half waves. This describes the motion of the hydrogen atom mathematically in terms of waves. Thus:

(5-40): 137/Cos (360/274)=137/0.999737 =137.036028

The relationship between G and the electrical permittivity constant becomes:

(5-41): G = 16 pi e / [e$_o$ C^2 $(137.036)^3$]

This calculates to be:

(5-42): G = 6.67224E-11

We can also relate Zo and GC, thus:

(5-43): Zo = [GC $(137.036)^3$] / 16 pi e

36

GC calculates to be:

(5-44): GC = 2.000284E-2 = 1/ 50 ohms

(5-45): Zo = 376.731 met³/cou sec² (ohms)

We see that the term GC is an admittance of approximately 1/50 ohms. We can also write an equation for h in terms of GC and Q^2.

(5-46): h = GCQ² [(137.036)⁴] / 32 pi e

This calculates to be

(5-47): h = 6.62611E-34

We can also add our standard electrical equation:

(5-48): e_0 Uo = 1/ C²

By feeding the equations into each other using the standard formulas of physics, we can derive all the above equations from the Sister 1 table of units. The net result is that the Sister 1 solution enables us to have a whole set of interlocking equations that relate h, c, Q, Uo, e_0, Zo, and G. The Sister 1 solution enables all the main constants of the universe to be inter-related. This was not accomplished using the other solutions.

Section 5-4: The Light speed Equations

For the Sister 1 solution we have units of meters cubed over seconds cubed and meters squared over seconds squared and various combinations of powers of meters and seconds. What do these units mean? In this section we will investigate the meaning of such complex units.

When the radius of the universe increases, both the charge and the energy of the universe drop. Looking back in time, we see that as we compress space-time, we charge up the universe and give it energy. We can look at energy from Equation 5-22 as charges moving in a space time volume. Thus:

(5-49): Energy (E) = Q C^3

In equation 5-49 we see that energy is charge times the speed of light cubed. In addition:

(5-50): E= M C^2

In Equation 5-50 we have Einstein' famous equation and energy is mass times velocity squared. Also from equation 5-49 we see that energy is charge time velocity cubed. Thus solving for M we get:

(5-51): M = QC

Equations 5-49 and 5-51 are the missing piece of the puzzle that Einstein started to solve. For momentum the equation is:

(5-52): MV = QVC

Finally for the momentum of a photon, the equation is:

(5-53): Mo C = Q C^2

In Equation 5-51 we find that mass in coulomb meters per second is a first order linear function of the light speed. In Equation 5-52 we see that momentum in coulomb meters squared per seconds squared is a planar function of light speed. Finally in Equation 5-49 energy in coulomb meters cubed per seconds cubed is a volumetric function of light speed. These equations permit the reader to see how the physics of the light speed Co and Cs dimensions work. They also enable a look at other possible coexisting universes which operate at different dual light speeds.

The above equations are the primary equations of the universe. Einstein was able to present one primary equation (Equation 5-50). This book presents the rest.

Section 5-5: The Bohr Expansion Velocity

The early universe expanded rapidly as the Cs/Co universe compressed at a high velocity near Cs and flowed into a small spherical surface at the big bang inversion. At that time, the ruler was very small and the time clock ran very fast. Scientists call this point a pinpoint. To us it looks like that but to an observer at the big bang it was somewhat large to the very tiny observer.

The pinpoint itself has zero size from the Co to the Cs dimension. In addition the wormhole between the dimensions has basically a zero diameter. Therefore the passageway between the dimensions looks like a pinpoint.

At the big bang inversion the expansion velocity reached the geometric mean of light speed Co and Cs. Thus:

(5-54) Expansion velocity = $(Co\ Cs)^{0.5}$

Since Cs = 18833Co, the geometric mean of the expansion velocity is:

(5-55) Expansion velocity = 137.23Co

This is the geometric mean of the expansion velocity. The exact solution to this transient problem is for future scientists and mathematicians to develop. This is just a ballpark number from an engineering perspective.

The universe stabilized as an expanding sphere moving close to light speed Co. At the same time another expanding sphere was moving at light speed Cs. The Co/Cs universe had formed but right away it started to radiate another sphere of light speed Cs/Co energy.

A time was reached when the universe stabilized and when the hydrogen atom was produced. After that happened, the Bohr atom radiated Co/Cs dot-waves which inverted into Cs/Co dot-waves. This caused the Bohr radius to increase in size.

We can then calculate the Bohr expansion velocity using electrical theory and the mass to charge conversion charts in Section 5-2. This will enable us to find the time of the universe since big bang. It will also enable us to find the wavelength of the dot-waves from the radius of the universe.

The exact constants within the equations are not readily determined. All that can be done from a ballpark analysis is to use the astronomical data and then fit the constants into the equations.

This effort gives us an equation that defines the exact relationship between the electrical world and the mechanical world.

We can produce a general electrical equation from magnetic field attraction theory. The exact constant such as pi, 4pi, 0.866, depends upon the geometry involved. It is necessary to find the best electrical fit that produces a universe in the order of 13.7 billion years as per the astronomical measurements. The Hubble telescope has improved the accuracy of the time estimate greatly since when I started this effort in 1981. This method will provide us with the means of deriving the time of the universe since the big bang from a linear perspective. If we use the actual number of oscillations of a hydrogen atom since after the big bang, the age of the universe approaches infinity by that ultra-fast time clock.

Years of studying the constants of the universe by numerical analysis and a simple engineering calculator has demonstrated to me that the universe is tied together by the constants 0.866, 1.414, e, pi, 4, 16, 137, and 137.036. These numbers always appear in various formulas.

Numerical analysis shows us that the above factors enter into many equations of modern physics. I accept such constants as being appropriate.

In general from an electrical analogy perspective the gravitational force is driven by the discharge of the dot-wave charges throughout the universe. As the universe expands dot-wave current flow of both positive and negative nature extends from all matter and photons outward to the radius of the universe. This causes spherical magnetic type fields that are constantly expanding.

The electrical analogy helps to explain the space time force. No doubt Einstein would have looked at the problem from his space-time perspective. Yet as an Electrical Engineer the electric analogy is much easier for me to understand.

This force can also be viewed as the contracting force acting upon the expanding Bohr Orbit when we view the universe from the mechanical perspective. Therefore we can write the gravitational field equations in terms of the expansion of the Bohr orbit.

At the same time as the universe expands, the Bohr orbit expands and this causes a force between the present orbit in light speed Co and the past complete orbit in light speed Co. Likewise a similar force exists between the past proton inner oscillation and the present inner oscillation. They can all be looked at as similar to Ampere's laws current flow problems. The force between past and the present tends hold back the proton as it radiates dot-waves and expands.

The hydrogen atom becomes the standard because there are simple coulomb attraction forces and gravitational forces as well. Since the Bohr atom is expanding, the protons are also expanding and the neutron radius is expanding as well. The general expansion of the universe common mode produces the decay of charge all over the universe and the general gravitational field. This can be viewed as the loss of mass within the particles or the loss of charge within the particles.

The gravitational force of concern is the force between two hydrogen atoms. This is the same force that operates upon two heavy metal balls in a lab. Likewise it is the same force which holds us to the Earth.

Let us look at the force between two hydrogen atoms by producing a magnetic attraction equation that relates the spinning electron of one atom interacting with the Bohr Orbit expansion of the second atom. One term will be large and the other small as shown in the following equation:

(5-56):

$$GM_HM_H/R^2 = 2Uo\ (QC/137.036)[4pi\ QV_B{}^*\ ]\ Cos30^\circ/(R^2)$$

Equation 5-56 is a general electrical equation relating a gravitational force between two hydrogen atoms to the interaction of the spinning electron in the first Bohr Orbit of one atom producing one magnetic field represented by the term $QC/137.036$, and a magnetic image caused by the slow expansion of the Bohr radius of the second atom as it radiates dot-waves as represented by $4pi\ QV_B{}^*$.

The factor (2) is caused by the electron of atom 1 reacting with the field of atom 2 and the electron of atom 2 reacting with the field of atom 1. This doubles the force.

There is a vector angle between the Bohr orbit electrical image of the present and the radiated image of the dot-

42

waves of the past. A typical electrical vector angle of 30 degrees has been used. This angle is very geometric and also brings the age of the universe calculation into alignment with the astronomical data. However, the exact angle is for future mathematicians to obtain. The radiated image must lag the present image as it aligns to it. The present image would be the leader while the prior radiated image would have to align itself to the leading image.

The velocity $V_B{}^*$ represents the motion of the Bohr radius as it expands slowly in time.

The constant G represents the gravitational constant of 6.67260E-11, M_H represents the mass of the hydrogen atom of 1.67353E-27, Uo is the electrical permeability of 1.25664E-6, C is the speed of light of 2.99792E8, Q is the charge of both the electron and the proton of 1.60218E-19, and R is the distance between the two atoms.

The equation for this electrical analogy form of the gravitational force can be interpreted in two different ways. If we know the gravitational constant and the phase angle then we can obtain the time of the universe since after the big bang from the expansion velocity $V_B{}^*$. Likewise if we know the expansion time of the Universe from astronomical measurements we can obtain the constant G and the phase angle by working backwards.

From equation 5-56, we can solve for the velocity of expansion $V_B{}^*$ of the Bohr orbit.

(5-57): $V_B{}^* = 1.21667E-28$ meters per second

The standard physics equation for the Bohr radius is:

(5-58): $R_{Bohr} = 137.036$ h / (2pi M_E C)

(5-59): $R_{Bohr} = 5.29178E-11$

43

The Bohr radius calculates to be 5.29178E-11 where M_E = 9.10939E-31, h=6.62608E-34, and C= 2.99792E8 for the Bohr orbit.

The time of the Universe is:

(5-60): $T_u = R_{Bohr} / V_B^* = 4.34940E17$ seconds

(5-61): $T_u = 13.7827$ billion years.

For this calculation 365.242 is used for the number of days per year.

This method is within 0.58 percent of the astronomical data of 13.7 billion years plus or minus 0.1 billion years. This would indicate that the choice of 30 degrees for the vector angle is reasonable. It also shows that the electrical analogy equation provides a reasonable answer. Of course future mathematicians will produce more accurate results.

If we write equation 5-41 in terms of Uo, we get:

(5-62): $G = 16$ pi e Uo/ $(137.036)^3 = 6.67223E-11$

This expression gives G (6.67260E-11) to an accuracy of 0.005 percent. Here G is clearly shown as the electrical permeability times the constant:

(5-63): Constant = 16 pi e / $(137.036)^3$

This constant appears to be a conversion between a wave (16 pi e) and a Bohr radius moving in three different rotations for the X, Y, and Z axis. The exact mathematical reasoning is beyond my engineering abilities. However this type of number has always appeared during my numerical analysis of the proton and the neutron.

The accuracy of the time of the universe by my hydrogen atom equation will vary due to differences between the hydrogen atom, the neutron, and complex structures of

44

molecules in the universe including black holes and neutron stars.

In addition, the expansion of the universe is moving from a high of 137.23Co toward a low of Co. Thus the expansion is quite non-linear and this is just a best fit engineering approximation to very complex mathematics.

The above method says that the complex dot-wave radiation acts like a simple electrical attraction problem. The two hydrogen atoms exist within the gravitational field which pushes them together. Within the field the atoms behave as if they attracted each other whereas it is the radiated field that is pushing them together.

Section 5-6: The Radius of the Universe

After the big bang inversion and linear stability, the universe has been expanding a little faster than Co. Initially it was expanding very rapidly and this stabilized the universe as an expanding spherical plane.

The distance from the initial point of stability after the big bang point to the present took 13.8727 billion years or 4.34940E17 seconds.

The radius of the universe Ru is:

(5-64): Ru = Tu C

In Equation 5-64, the radius of the universe is the time of the Bohr expansion times the speed of light Co.

(5-65): Ru = 1.30392E26 meters

In equation 5-64 the radius of the universe is 1.303926E26 meters for a straight line linear analysis. As the universe radiates dot-waves, it expands exponentially as it loses energy exponentially. Both the ruler and time clock expands. The exponential function enables us to

45

look at the very small universe and the very large universe from a linear analysis. Thus the universe of the much smaller size long ago looks the same to a neutral observer then as of today. The measurements of the size of a proton from long ago would appear the same as today since as the proton expanded, the ruler expanded as well.

The radius of the universe in the Cs dimension is:

(5-66): $Rus = Ruo\ (Cs/Co)$

(5-67): $Rus = 18833Rus$

(5-68): $Rus = 2.45567E30$ meters

The outer band of our universe is 2.45567E30 meters. Thus we live inside a huge outer universe. In addition it keeps growing larger and larger every day.

Section 5-7: The mass of the particles in the Cs dimension

The mass of the electron, proton, and neutron in the Cs dimension is the square of the ratio Co/Cs times the mass of the particle.

(5-69): $M_E = 2.56838E-39$

(5-70): $M_P = 4.71593E-36$

(5-71): $M_N = 4.72244E-36$

Section 5-6: The Mass of the Universe

Although there are various formulas that could be used to calculate the mass of the universe at our light speed Co, all require correlation with the astronomical data. Thus it is best to use the work of Eddington for the equivalent number of neutrons in the visible universe.

(5-72): $N_N = 1.586768E79$

Since the mass of the neutron is 1.67493E-27Kg, the mass of the visible universe is:

(5-73): Mu = 2.65773E52Kg

According to the astrophysicists, the mass of the visible universe is only 4% of the total mass/energy of the universe.

The dot-waves within space make up a large part of the Co/Cs mass/energy. Thus the gravitational and electromagnetic fields are composed of oscillating neutral dot-waves, oscillating positive charged dot-waves, and oscillating negative charged dot-waves.

In addition there are Cs/Co high velocity dot-waves of the three different types. As long as the high velocity dot-waves remain within our Co/Cs radius of the universe they will affect us.

The total equivalent mass of the universe is 25 times Mu. Thus:

(5-74) Mu (total) = 6.644325E53 Kg.

This same amount of mass has a total equivalent mass in the Cs universe of:

(5-75) Mu (total Cs universe) = 187336E45 Kg

Section 5-8: The Mass and Charge of a Dot-wave

We know the radius of the universe. We want to know the mass of the dot-wave. Once we know this we can calculate the number of dot-waves in an electron. The electron is composed of only negative dot-waves. This will enable us to find the charge of a dot-wave.

The structure of the universe is such that every point on the physical surface plane of the universe is at a distance

Ru from both the absolute center of the universe and the outer sphere of the universe at a distance 2Ru from the absolute center. This is illustrated in the picture of the universe as shown on the page before Chapter 1.

We have a spherical surface at radius 2Ru from the center of the universe and at a distance Ru from us. Ru appears as the primary variable of the universe. No matter how the universe expands or contracts this variable will be determined by the driving force of the universe.

It is self-evident to me that this variable surface looks like the charge Q. At the big bang Q was very large. As the universe expanded it got smaller and smaller. Although the distances are huge now, the exponential function causes the universe of today to look like the universe of yesterday and the universe of tomorrow. Therefore we can write simple electrical equations as if the long distances of Ru today operated the same as yesterday.

The individual dot-waves which are not compressed as part of electron, protons, sub-particles, etc. oscillate from a zero point to the radius of the universe. Space has some individual dot-waves whereas most things are composed of huge numbers of dot-waves. The capacitance of the individual dot-waves is:

(5-76) $C_{DW} = 4$ pi e_o Ru

In Equation 5-76 the capacitance of an individual dot-wave which is a sphere equal to the radius of the universe and a zero point on the surface of the universe where it oscillates to the light speed Cs dimension.

The voltage of the oscillating dot-wave is:

(5-77) $V = Q/ 4$ pi e_o Ru

The inductance of the dot-wave is

48

(5-78) $L_{DW} = 4 \text{ pi } U_o \text{ Ru}$

As the Dot-wave oscillates from r= 0 to r= Ru, the energy will flow from the capacitance of space to the inductance of space. The inductance and capacitance energy will be equal. Therefore the total energy is:

(5-79) $E_{DW} = C_{DW} V^2$

The total energy is the capacitance of the dot-wave and the voltage squared. Thus

(5-80) $E_{DW} = Q^2 / 4 \text{ pi } e_o \text{ Ru}$

The energy of the dot-wave is the square of charge divided by 4 pi, and divided by the permittivity, and divided by the radius of the universe.

(5-81) $E_{DW} = 1.76935 \text{ E-54}$

The mass of a dot-wave is the energy of a dot-wave divided by the speed of light squared.

(5-82) $M_{DW} = E_{DW} / Co^2$

(5-83) $M_{DW} = 1.96867 \text{ E-71 Kg.}$

In my prior work using a simple DC analysis I got 1.5662 E-72Kg. The difference equals:

(5-84) Ratio of the AC analysis to the DC analysis = 4 pi.

I wasn't certain that the 4 pi was correct for the DC analysis. This method appears more accurate. The actual mass of the dot-wave does not affect the theory. It merely changes how many dot-waves are in a particle or photon. However future mathematicians will most likely produce better equations with more accurate results. My work is just an engineering approximation so that future scientists and mathematicians can continue this work with superior results.

We can now find the charge of a dot-wave. The electron only contains minus dot-waves. The proton has many bipolar dot-waves and an excess of positive dot-waves equal to the number of negative dot-waves within the electron.

The number of dot-waves in an electron is:

(5-85) $\#E_{DW} = M_E / M_{DW} = 4.62718E40$

Where $M_E = 0.910939E-30Kg$

The charge of a dot-wave is:

(5-86) $Q_D = 1.60618E-19/4.62718E40$

(5-87) $Q_D = 3.47119E-60$ Coulombs

The number of dot-waves in the proton is

(5-88) $\#P_{DW} = M_P / M_{DW} = 8.49619E43$

Where $M_P = 1.67262E-27$ Kg

The number of positive dot-waves for the proton is:

(5-89) $\#P_{PDW} = 4.62718E40$

The number of the sum equal amounts of positive and negative Dot waves for the proton is:

(5-90)$\#P_{2BPDW} = 8.49619E43 - 4.62718E40 = 8.49156E43$

The number of bipolar Dot-waves within the proton is:

(5-91) $\#P_{BPDW} = 4.24578E43$

The number of Dot-waves for the neutron is:

(5-92) $\# N_{DW} = M_N / M_{DW} = 8.50793E43$

Where $M_N = 1.67493E-27$ Kg.

According to Quantum physicists, there are three slightly different neutrinos. Experiments at CERN have shown the possibility that some are traveling at speeds greater the Co. One possibility is that the neutrinos are composed of Cs/Co energy and when they transition from the light speed Cs dimension, they spend a little more time at high speeds than the normal minimum time for everything else. That is left for future scientists to consider.

The universe we live in has a mixture of Co/Cs dot-waves and Cs/Co dot-waves. These appear as particles and photons. We than have photons which travel at Co and photons which travel at Cs.

The Cs photonic energy helps tie the universe together in information. This produces some strange space time effects. A star could explode millions of light years away from here. We see the exploding galaxy but Cs photons arrived here before the galaxy exploded.

The net result is that if we have an instrument which was sensitive to Cs photons, we could observe what the star looked like before it exploded.

The question of time travel comes up. If we could send data at light speed Cs, it would be traveling much faster than Co. It would seem that we have gone backwards before the event. However what defines the time of an event? Einstein would define an event at light speed Co. Then if you went faster than Co, you would go back in time. However an event actually occurs based upon light speed Cs.

We see the event based upon Co which is much slower than the actual time. We cannot go back in time to the event which produced both Co and Cs photons. Therefore we cannot change the event itself. All we can do is observe the event in light speed Cs photons and electrical waves first before the light speed Co photons and electrical waves arrive. In order to do that we have to record and store the information of the universe from light speed Cs data and then find a star that exploded. Then we have to rewind the light speed Cs data to see the star prior to its explosion. We can only look back in time if we recorded the event on a memory device.

The number of dot-waves in the visible universe is the mass of the visible universe divided by the mass of each dot-wave. Since from Equation 5-73, Mu =2.65773E52Kg, the number of dot-waves in the universe is:

(5-93): $N_u = M_u / M_D$

(5-94: $N_u = 1.35001E123$

Where $M_D = 1.96867 E-71$

In equation 5-94 we have broken the visible universe down into a large amount of dot-waves of electromagnetic energy quanta by the normalization process. At this point in time the entire model universe has been quantized into tiny bits and pieces of energy and charge. When we add in the invisible universe we get approximately 25 times as much.

Chapter 6: The Dual Light-Speed Dot-wave Theory

Section 6-0: Introduction

The Dual Light-Speed Dot-Wave Theory states that the entire universe is a huge dual light-speed electromagnetic-photonic-gravitational field. It states that everything in the universe is electrical, photonic, and gravitational in nature. It states that the property of mass is due to an oscillation of the dot-waves from the light speed C_O dimension to the light speed Cs dimension.

The dot-wave oscillation is such that a light speed Co spherical wave flows into a pinpoint at the junction of the Co/Cs dimensions. The oscillation disappears from the Co dimension. It emerges within the light speed Cs dimension where it expands and then contracts to enter a new pinpoint at the junction of the Cs/Co dimensions. It disappears from the Cs dimension and passes through the new pinpoint where it emerges within the light speed Co dimension again. In the process the wave goes from a high mass low light speed configuration to a low mass high light speed configuration. The energy is constant for the dot-wave on both sides of the barrier.

When the wave returns to the light speed Co dimension it has shifted its position depending upon how fast the dot-wave was traveling with respect to the speed of light Co. To an observer observing the dot-wave from the Co dimension, it would appear that the space has shrunk.

Einstein would say that the distance shrunk and the time clock slowed. Mathematically that is a good approximation to what happened. His math is very good but it is not true. Space did not shrink. The time clock did not slow. There was no legitimate time clock.

We went from a light speed Co dimension and traveled very fast in the light speed Cs dimension. Since the distance we traveled was much smaller at the high light speed dimension, the time clock could only do a small percentage of its ticking cycle.

Einstein's analysis gives the correct answers but the physics is not true. He has presented his theory based upon a single light speed solution. Thus his work is just a partial model of reality.

The original year 2000 Dot-wave theory states that mass can be represented as a property of charge, space, and time. In a similar fashion charge can be represented as a property of mass, space, and time. This enables us to find the variation of the various constants of the universe as the universe expands from a pinpoint toward maximum dimensions.

The Dot-wave theory states that everything in the physical universe can be represented by a multitude of only three basic types of things. For the electrical viewpoint, there is a multiplicity of plus dot-waves, minus dot-waves and bi-polar dot-waves. The electron, the proton, the neutron, and the photon are all composed of dot-waves. The same is true of all the sub-particles. The only difference between things is the particular energy configurations that each particle or sub-particle possesses. There are nine basic things and many variations of things which possess different levels of the nine basic things. This can produce a huge amount of particles and sub-particles.

The Dot-wave theory states that the electron is composed of minus dot-waves equal to the total charge Q= 1.602E-19 coulombs. The proton contains approximately 1836 times as many dot-waves as the electron but has an excess of positive verses negative dot-waves equal to the charge Q= 1.602E-19 coulombs.

The dot-wave is a quantum of the electromagnetic/ gravitational field. It is a spherical and or angular and or linear energy form which comes in plus charges, minus charges, and neutral plus/minus photonic energy. In addition the dot-wave oscillates from a low light speed high mass configuration to a high light speed low mass configuration.

During the oscillation, the dot-wave readily changes energy forms as it interacts with matter and the gravitational field. It can be a linear photon one minute and become a spherical wave another minute. When the dot-wave is part of a photon traveling at the speed of light Co, it no longer oscillates in a somewhat spherical pattern to create mass. Since the photon travels at Co, it will never return to any point that it passed either at light speed Co or at light speed Cs.

The electron can have a spherical shape and look like a particle. If we add photons to it and bring it toward the speed of light Co it approaches an electro-photon. Thus you have a negatively charged photon. This photon will travel almost at the speed of light. If you keep adding photons, it will reach very close to the speed of light and look like a line of charge.

The electron can only achieve a speed of slightly less than Co. The mass will increase as the speed increases and the energy of the magnetic field will increase as well. It should spin faster as it reaches near light speed. However it will still maintain a partial spherical energy pattern and therefore it will always have mass.

Normally an electron will look like a particle and a wave. Within the electron the individual dot-waves are oscillating from light speed Co to light speed Cs. The percentage of dot-waves reaching light speed Cs at the same time depends upon probability and the particular experiment. If all the dot-waves become synchronized, the electron could escape its present location and travel huge distances to become an electron elsewhere

Due to the more complex structure of the proton the probability that the proton will end up with fully synchronized dot-waves is small as compared to the electron. Likewise the probability of an entire spaceship having fully synchronized dot-waves and being able to travel at speeds near light speed Cs is basically zero.

The question is how we can calculate the long term properties of the gravitational field and the electromagnetic field in a simple manner. The electro-scientists and mathematicians have produced various electrical models which work very well. Thus we can understand to a great extent how gravity and electromagnetism works from the inversion at the big bang to maximum expansion using electrical models.

Then we have to add the properties of oscillating dot-waves to the mix. For this effort, I must rely on engineering models. The models do not have to be perfect but they have to explain how gravity and space and time work over time.

We will now enter the world of the dot-wave theory. We have already studied the mass to charge velocity conversion in in the prior chapter. Now we can expand on that chapter to include some additional points.

For a constant energy universe, the energy of each dot-wave is a constant. The same is true of the charge of a dot-wave. The charge Q of both the electron and proton drops as the universe expands. At the same time, the mass of the proton and electron drops as well.

The charge Q of the electron and the mass of the electron after big bang was much larger than today. In the future it will be even less. The reason is that the universe radiates dot-waves. The means of the radiation is the oscillation of the dot-waves from the light speed Co dimension to the light speed Cs dimension. Some of the dot-waves remain in the Cs dimension. They become Cs/Co dot-waves.

In effect the universe is returning to a high light speed low mass configuration. The same is true of the photons. They too radiate away slowly. The mass of the universe drops as the universe radiates away. This radiation causes the force of gravity upon every particle in the universe.

Section 6-1 Dot-wave Current Flow

When we looked at the hydrogen atom in terms of an electrical circuit, we noticed that it expanded in time. The hydrogen atom right after the big bang was much smaller than the hydrogen atom today. The charge Q was much larger and the mass was much larger.

The electron was radiating negative dot-waves. This caused an outward negative current flow. At the same time, the proton was radiating both positive and negative dot-waves. Since there are more positive dot-waves in the proton than negative dot-waves, the proton radiates more mass than the electron but an equal amount of net positive electrical current.

This causes the charge to mass ratio of the proton to remain a constant. Therefore the smaller universe right after the big bang looked very similar to the universe of today. Of course galaxies and stars were forming but the basic physics was very similar.

The dot-wave charge and the dot-wave mass are invariant but the charge Q of the electron and proton keeps dropping as the universe expands. The same is true of the charge Q at the radius 2Ru.

Section 6-2: The Model Universe

Since the universe is radiating dot-waves we can look at the universe as an electrical engineering problem. Let us start out with some very simple algebraic equations as we search for an understanding of the relationships between the electrical world and the mechanical world of the universe. In order to understand how the universe functions, it is necessary to start out with a simple model of the universe and then explore the model and expand upon it until it matches the actual measurements obtained in the laboratory. The universe is modeled upon huge numbers of expanding charged spheres of energy.

For our model of the universe, every point in the universe is approximately equidistant from the center of the universe. All electromagnetic/gravitational fields within the universe form an outer spherical shape such that all fields from the surface of the material universe completely fill all the spaces on the outer spherical surface of the universe. The spheres all interlock in three dimensions of distance such that the invisible physical light speed Co/Cs universe always looks like an absolutely perfect sphere of radius 2Ru when viewed from outside the universe. However from any point on the surface of the visible universe, the universe basically looks like a sphere of radius Ru from that point.

After the spherical surface of radius 2Ru lays the Cs/Co dimension. This extends outward to 18833Ru at the present time. As the Co/Cs outside radius expands into the future, the Cs/Co radius will reach close to infinity. However a point will be reached where most Co/Cs dot waves have been converted into Cs/Co dot-waves and the expansion will stop. Then the universe will contract and the next big bang will be set up.

To understand what the physical universe looks like we can draw the universe with a compass. We can look at the universe as a spherical surface. We live upon a plane at a distance Ru from the center of the total universe. If we draw a circle of radius Ru on a piece of paper this will represent a two dimensional view of the universe. We live on the circumference of the circle. We can then take a compass and set its radius to Ru. After that we can draw a huge number of circles of radius Ru all around the circumference. The net result is the formation of another complex circular structure of twice the radius as the original circle. This is shown in the picture of the universe prior to the start of the first chapter.

Every point on the original circle is basically at the center of its own circle. All these circles form a dark spot in the absolute center of the universe. The net result is that every point in our universe is equidistant from the absolute center of the universe and the outer circumference of the circle produced by the huge number of circles centered upon the original circle. If you spin the entire universe it will turn into an ellipsoid. The actual shape of the physical universe can vary over time. As the visible universe slows in rotation, it will look more and more like a simple sphere. The non-visible universe however will always form a perfect sphere since as you move further and further away from the visible universe, the photonic dot-waves always form a sphere. Thus we live within a perfect sphere at radius 2Ru from the absolute center.

Although the universe consists of a complex pattern of galaxies approximately equidistant from each other, in order to find simple equations, we can look at a situation in which all energy existed at a single pinpoint and the explosion at big bang only caused one galaxy to exist with all the stars at a distance Ru from the center of the explosion. We can then look at a simplified model of the universe.

Let us look at a single sphere of photonic/gravitational energy expanding from a pinpoint as representing the universe. As the photonic energy expands, the radius of the universe also expands and a standard ruler and a standard time clock expand as well. Time clocks slow down and rulers stretch as the universe expands. Notice that if the universe expands at light speed, it expands much faster when it is small then when it is large. The rate of expansion or the acceleration of the photonic sphere is:

(6-1): $A_U = (C^2)/R_U$

Equation 6-1 states that an expanding sphere of photonic energy has acceleration, A_U of meters/second^2 that is simply the speed of light, C squared divided by the radius, R_U of the light sphere. An idealized universe at every point follows the form of equation (6-1)

This simple equation forms the basis of an idealized universe comprised electromagnetic and gravitational fields, which experiences time and distance elongation.

The total force acting upon the model universe is simply the mass of the Universe M_U times the acceleration of the universe A_U:

(6-2): $F_U = M_U A_U = (M_U C^2)/ R_U$

The universe will go from a point of maximum mass to a point of minimum mass within the visible light speed Co/Cs universe. At that point most light speed Co/Cs mass is used up and the visible universe stops expanding. Theoretically this can occur at infinity radius since distance expands with time and the clock slows with time. However if we use a normalized clock we can derive an equivalent cycle time of the universe.

When the non-visible light speed Co/Cs universe of 2Ru reaches maximum size very few physical particles or sub-particles exist.

The physical light speed Co/Cs universe at radius Ru from the absolute center will eventually be reduced to a minimum amount of energy. The same is true of all the Co/Cs energy that is left. It will end up on the spherical shell at absolute radius 2Ru. Both shells will continue to oscillate between light speed Co and light speed Cs. They will form the basis of the next Co/Cs universe after the next big bang.

The physical expanding universe will eventually be gone but the compressing universe will remain. At that time the Co/Cs universe will have no more input energy to continue to expand. In many respects the universe of Co/Cs light speed and energy will be like a balloon. We can then look at the oscillation of the universe in terms of the general gas law.

The ball of light speed Cs/Co energy at radius 2Ru will compress at light speed Cs/Co toward a pinpoint or small surface sphere. Once this happens the energy will invert and the Co/Cs dot-waves will reform.

The big bang is really a space time inversion of light speed Cs/Co energy being converted into light speed Co/Cs energy. High light speed low mass energy will be converted into low light speed high mass energy at the inversion

The particles and sub-particles that formed after the inversion were higher energy. The protons and electrons had both higher mass and higher charge. The photons also were higher energy photons. As the universe expanded the entire energy and charge levels dropped common mode. Measurements of a red photon one billion years ago would look the same as today when using a smaller ruler at that time.

The mass of a proton measured a billion years ago would also measure the same as today due to the shortened ruler at that time and corresponding measuring devices. This makes it difficult to detect the loss of mass/energy in the visible universe. However indirect measurements of light speed Co/Cs dark mass/energy make it obvious that approximately 96 percent of the visible universe has been erased.

The total light speed Co/Cs mass/energy at big bang was about 25 times what it is today. We can then assume that the universe radiated mass/energy back into dark mass/energy as a declining exponential function. Due to the high light speed during contraction the buildup of mass/energy from a maximum radius 2Ru to basically a tiny radius at Ru was a fast rising exponential function.

Thus the universe goes from 2Ru where Ru is near infinity to Ru where Ru is near zero. The expansion time is To when the light speed is Co/Cs and the contraction time is Ts when the light speed is Cs/Co.

All the while things were going on within the visible light speed Co/Cs universe. At that time mass was leaving our visible universe at light speed Co on the surface radius Ru and entering the invisible universe at light speed Cs. Since the mass of our universe was decreasing as a declining exponential function, this enables us to produce some simple equations for a ballpark engineering model of the universe. A more scientific model will be produced by future scientists and mathematicians as they learn more about the dual light speed universe.

In this book we look at things from both an electrical and a mechanical viewpoint. The physical universe can be described either from electrical equations or mechanical equations.

From a mechanical viewpoint, the force driving the expanding universe is the conversion of light speed C_o/C_s energy at the surface sphere Ru from the absolute center into light speed C_s/C_o energy. This expands both the physical universe on the surface plane Ru and the outer boundary of the universe at 2Ru. Mathematically this causes a loss of mass per unit time. The force acting upon the entire universe is:

(6-3): $F_U = -C_o \, d(M_U)/ \, d(T_U)$

In equation 6-3 we see that the force acting upon an expanding model universe is the light speed C_o times the derivative of the mass of the universe with respect to the time of the universe. Notice that the expanding bubble of the gravitational field has less and less force acting upon it as it expands more and more toward infinity since the mass is decreasing toward zero. Correspondingly for the electromagnetic field the charge is decreasing to zero as well.

Although the charge of protons and electrons decrease with time, the charge and mass of an individual dot-wave is constant at both light speed C_o and light speed C_s. The electron loses negative dot-waves. This causes both the mass and charge of the electron to drop in time. The same is true of the charge and mass of the sub-particles which make up the proton. Photons also lose dot-waves and over time higher energy photons turn redder. However since the universe operates upon a common mode exponential function, things look much the same over time.

From the astronomical data of dark matter/dark energy we can say that the mass/energy at the big bang was twenty five times higher than today. If we use a variable time clock the energy at the big bang inversion could have been near infinity. The exponential function helps us to understand the expanding universe. A normalized exponential function turns a variable time clock into an ordinary time clock which makes the equations simpler. Future mathematicians and scientists can produce more detailed equations. However the Engineering models suffice to explain the basics of the Dual Light-Speed Dot-Wave theory.

The radius of the universe R_U is related to the time of the Universe by the following:

(6-4): Ru = C Tu

The radius of the universe for a simple linear solution is simply the speed of light times the time of the universe since the pinpoint at big bang. The analysis looks at a universe which has already expanded at a very high light speed and has stabilized to about light speed Co. The initial expansion speed was most likely close to the geometric mean of Co and Cs. The simple model here is a ballpark model in order to produce an approximate Engineering model universe.

The energy of the visible universe is simply the mass of the universe times the speed of light squared. Thus

(6-5): $E_U = F_U R_U = M_U C^2$

Notice that all these equations are very simple general equations, which describe an expanding photonic sphere. Yet, they will produce reasonable results when the concepts are applied to the standard equations of physics for the various calculations shown in this book. The purpose of this Chapter is to produce idealized equations which look at how the universe actually works. These will never exactly match the way we see the universe from the point of view of particles and photons although things will be similar.

If we set equation (6-3) equal to equation (6-2) and solve

63

Using equation (6-4) we get:

(6-6): $M_U C^2/R_U = -C \, d(M_U)/d(T_U) = -C^2 \, d(M_U)/d(R_U)$

In addition since the net or differential driving function is the loss of mass as the universe expands:

(6-7): $M_U R_U$ = Constant (Kilogram Meters)

One solution to Equations 6-6 and 6-7 is an exponential function.

This simple solution to an expanding light sphere is

(6-8): $M_U = M_O \, e^{-x}$

(6-9): $T_U = T_O \, e^x$

(6-10): $R_U = R_O \, e^x$

(6-11): $C_U = C_O$

Where X is a driving function which varies from minus infinity to plus infinity.

In equation (6-8) we see that the mass of the universe decreases as the driving function (x) of the universe increases. In equation (6-9) we see that the time of the universe expands as the driving function (x) of the universe increases. Finally in equation (6-10) we see that the distance of the universe expands as the driving function of the universe expands. Thus as the photonic sphere increases, both the ruler and the clock expand for a constant light speed universe as per equation (6-11). The driving function can be a simple linear function. It could also be a more complex function.

The photonic mass/energy flow operates in the same manner as the electric field. Therefore the electric field also experiences the same type of flow. Therefore the charge Q of a proton or electron decreases at the same rate of decline as the mass of the proton or electron. Some variation may be possible but it is most likely that the charge to mass ratio of the electron and proton is constant over time.

Since both charge and mass decrease as the universe expands, the following equations also apply:

(6-12): $Q = Q_0 e^{-x}$

In equation 6-12 we see that charge decreases as an ordinary exponential. It gets smaller as both time and distance increase.

(6-13: $U_0 = U_{0o} e^{2x}$

In equation 6-13, the electrical permeability U_0 increases to the second power as both time and distance increase.

(6-14): $K = K_0 e^{2x}$

In equation 6-14, the Coulomb's constant K increases to the second power as both time and distance increase. Likewise for the electrical permittivity:

(6-15): $e_0 = 1/K = e_{0o} e^{-2x}$

In equation 6-15 we see that the electrical permittivity constant e_0 decreases as the universe expands. The speed of light is:

(6-16): $C = 1 / (U_0 e_0)^{1/2}$

In equation 6-16 we see that the speed of light remains constant as the universe expands since the electrical permeability constant increases while the electrical permittivity constant decreases. The impedance of the universe is:

(6-17): $Z_U = Z_{Uo} e^{2x}$

In equation 6-17 we see that as the universe goes toward infinity in size, the impedance of the universe also goes toward infinity and it becomes an open electrical circuit. When the light speed Cs/Co universe was compressed toward a pinpoint at minus infinity for a general model, the impedance of the forming universe was zero.

Prior to the inversion at big bang, the Co/Cs universe did not exist except for a subliminal image. As the Co/Cs universe expanded and changed into the Cs/Co universe, the very high impedance of the Co/Cs universe was also high for the Cs/Co universe.

As the universe compressed, the impedance Cs/Co universe went from a very high value to near zero. Then the inversion occurred and the impedance of the Co/Cs universe was basically an electrical short circuit.

From this analysis we see that the electrical universe model has an inductive/capacitive structure which varies from a short circuit at big bang to an open circuit at infinity.

From these simple relationships of an expanding photonic energy sphere, a model of the universe can be presented. It is necessary to add to the model specific constants, which will produce values that are close to what we see and measure.

When we look at the universe from a constant light speed Co/Cs model, the radius of the universe expands with time and the mass of the universe decreases with distance, therefore the gravitational constant G increases as the square of the radius of the universe. Since the mass decreases, the term GMM is a constant except for nonlinearities.

Let us look at the force between to two masses over time.

(6-18): $F = GM_1M_2/R^2$

As we look at equation 6-18 we see that the gravitational force decreases as the square of the distance as the universe expands. Since the GMM term is constant, the forces are getting less and less.

As we reach toward full expansion, the gravitational forces within stars and planets will weaken as the square of the radius of the universe. Therefore, stars and planets will explode. The protons will do no better since they experience a slowing of their internal oscillations. Thus protons and electrons experience their own red shift effect, as do photons. They will be destroyed as we head toward maximum expansion. This is called the mini-bang where the entire light speed Co/Cs universe returns to basic dot-wave energy.

Since the mass/energy of the invisible universe is about twenty five times more than the visible universe, we have both a common mode expansion of the universe and a fast expansion as protons, electrons, particles, and sub-particles radiated huge amounts of Co/Cs dot waves into space.

The dark matter/ dark energy make up the structure of the gravitational and electromagnetic fields. These are composed of low energy/ low charge dot-waves. In addition a lot of Cs/Co energy initially came directly from the Cs/Co electromagnetic and gravitational fields in the compressing Cs/Co invisible universe.

The net result is that the simple model here is merely a ballpark solution for a very complex problem which would require large teams and huge computers to solve.

The exact speed of rotation of the universe at the inversion may have been very large. Over time it slowed and as we head toward infinity it will be at zero speed. The exact details are left for the future mathematicians to work out.

Section 6-3: The Dot-wave Oscillating Universe

The compression of the dot-waves produced the particles and sub-particles during the period of mini bangs after the initial big bang. This resulted in a very short period of intense non-linear radiation followed by a long period of linear exponential radiation. The universe could then slowly expand toward infinity.

Ever since the big bang occurred, the universe has been radiating dot-waves into both the visible universe of light speed Co/Cs and the invisible universe of light speed Cs/Co. The visible matter/energy of the visible universe at radius Ru has been disappearing. At the same time, the energy at the outer shall at radius 2Ru has been increasing. A point will be reached where the visible matter and energy of the universe will be gone. At maximum radius the time clock of the visible universe will stop. At that point we will only have the memory of this universe in the outer shell at radius 2Ru of light speed Cs/Co.

At full expansion the total energy of the universe has not changed. The universe will become an extremely high light speed invisible universe. Then the universe will start to rapidly compress again.

The universe will then oscillate forever following a variable time exponential function. When the universe increases in size, it starts off at a radius of zero and quickly rises at a light speed which varies from Cs to Co with a geometric mean of:

(6-19) Geometric mean Rising light speed = $(Co \times Cs)^{0.5}$

In equation 6-19, the geometric mean is the square root of the product of light speed Co and light speed Cs.

Finally the universe slows to a light speed close to the speed of light Co. This will continue until the visible physical universe is gone. At that time the outer shell will be oscillating between Cs and Co. There will be individual dot-waves of Co/Cs and Cs/Co from the radius of zero to the outer radius at 2Ru where Ru is a very large radius at that time.

The universe will then compress at a rapid speed with a geometric mean equal to the rising light speed shown by equation (6-19). During the next cycle is it possible that we will have an anti-matter universe at the inversion point. Protons will be anti-protons and electrons will be positrons. However the universe will operate the same way as today.

The question is how can we best find the cycle time?

The variable light speed exponential solution does not produce a sinusoidal shape but it has sinusoidal

68

components. Mathematicians could then produce curves of the most likely cycle time based upon a time variable exponential curve.

That could be done in the future but for an engineering solution for the cycle time of the universe we can consider that we get an oscillation based upon the inductance and capacitance of space. In this way we turn a complex problem into one that can be solved by an engineering mathematical describing function.

Even though time and distance are expanding we could use a normalized time based upon our present measurements. This would give us a relative answer where the time clock appears unchanged.

Using this simple method, the oscillation of space-time is caused by the inductance and capacitance of a universe of radius Ru. The inductance equals the electrical permeability times the area divided by the distance. The surface area of the visible universes is $4 \pi Ru^2$ and the distance is Ru. Thus:

(6-20) $L_u = 4 \pi Uo Ru$

The capacitance is the permittivity e_0 times the area divided by the distance. Thus:

(6-21) $C_u = 4 \pi e_0 Ru$

The cycle time of the universe is:

(6-22) $Tc = 2 \pi (L_D C_D)^{1/2}$

(6-23) $Tc = 8 \pi^2 Tu$

Using Tu = 13.78 billion years, the cycle time of the universe would be

(6-24) $Tc = 1088$ billion years

This calculation is based upon our time clock. In reality as the universe stretches out toward infinity, the clock slows toward zero speed. As the clock slows the 1088 billion years becomes infinite time by our clock. If we used a variable speed clock then that clock would measure 1088 billion years.

This method turns a variable time exponential function into a simple electrical oscillator. This gives the person with an engineering level or technician level understanding the ability to see how the universe works.

The mathematicians and scientists require a greater more mathematical understanding. Therefore this work is only the start of the future understanding of the dual light speed universe.

Chapter 7 Linear Doppler Space Time

Section 7-1 Introduction

The Michelson/Morley experiment had several solutions which were debated long ago. One solution specified that the measuring instrument's dimensions varied as the Earth moved toward the sun or away from the sun. Another solution specified that the speed of light is constant regardless of whether a measuring instrument is moving toward the source of light or away from the source of light. Einstein took this solution and it became the basis of his special relativity. The debate still continues and equations can be written and explained several different ways.

The dual light speed universe theory specifies that the speed of light Co is invariant. The same is true of light speed Cs. These are constants. Yet how do rulers and time clocks vary. Einstein's equations work quite well so whatever equations are used, the answers must be identical with Einstein.

In my "Doppler Space time" book, I took the position that the answer was derived from the Doppler Equations. For this answer the root mean square of the Doppler Equations equals Einstein's equations. The Doppler equations have differences in the front of motion and the rear of motion but the geometric mean is the same as Einstein's equations.

Einstein used different reference platforms which move at the velocity V with respect to each other. Space and time react accordingly. Einstein has a time dimension which can shrink or expand. However there is no time dimension because time is only a measurement. You cannot go back in time because the events of the past are gone. They no longer exist except in photonic memory.

71

The photon travels at the speed of light Co because that is the speed that it moves when it comes from the light speed Cs dimension and returns to the light speed Co dimension. No matter where the photon is, the speed will always be Co.

Let us look at an object is moving with a velocity V with respect to us. It emits a photon of light. The photon leaves the object and the photon is moving at Co with respect to the object. The photon oscillates between the Co and Cs dimensions. It then encounters a receiver. The receiver is made up of dot-waves which switch from the Co to the Cs dimensions.

When the photon reaches the receiver it is traveling at Co with respect to the receiver. It is not possible for the photon to travel at a different speed than Co. Einstein would say that space shrunk as the transmitter and receiver moved toward each other with velocity V.

This appears to be true based upon measurement data. Yet space does not shrink and time does not elongate. Mathematically, Einstein's answers are quite correct. A mathematical alternative is found using simple Doppler Equations.

In this book for the dual light speed universe photons do not flow from the sun to the Earth based upon the light speed Co dimension alone. If we look at photons flowing from the sun to the Earth unchanged then we can argue for or against Einstein's special relativity. However once we realize that the photons oscillate between the light speed Co dimension and the light speed Cs dimension, it is easy to understand that the time it takes to move from transmitter to receiver will vary.

As the photons move along they will gain or lose energy from the gravitational field intensity at the junction points they occupy. If the transmitter and receiver are moving toward each other, the gravitational field intensity will be increased and the photon will turn toward the blue as it gains energy. If the transmitter and receiver are moving away from each other, the photon will turn toward the red as it loses energy.

The gain or loss of energy occurs at the transition point between the dimensions. In some respects these transition points could be considered mini worm holes. The photon leaves one dimension or universe and enters another dimension or universe.

A simple analogy is that the photons breathe. The photon inhales and exhales dot-waves. When the gravitational field density increases at a transition point, the photon will add dot-waves. When the gravitational field density decreases at a transition point, the photon will shed dot-waves.

When two gravitational fields are moving toward each other, the gravitational field intensity increases and a photon will gain dot-waves. This is the same effect as when a photon travels toward a star and the gravitational field keeps increasing in intensity. The photon will keep getting bluer as the energy of the photon increases.

The light is blue near the surface of the sun. This light comes from the radiation of photons from the atoms of the sun which are under a strong gravitational field. Once the photons leave the surface of the sun, the photons lose dot-waves as they breathe.

Photons are moving dot-wave patterns which are electrically balanced from charge perspective. Thus there are an equal number of plus dot-waves and minus dot-waves within the photon. The photon oscillates from the Co dimension to the Cs dimension but the oscillation is synchronized.

We can look at photons similar to a series configuration of negative dot-waves and an equal amount of positive dot-waves. When the negative dot-waves are moving at Co in the Co dimension, the positive dot-waves are in the Cs dimension. The positive dot-waves are moving in a similar manner to a standing wave except that it is phase locked with the negative dot-waves waveshape. Thus it has a net linear motion of Co with respect to the Co dimension.

Then the negative dot-waves move into the Cs dimension. At the same time the positive dot-waves move into the Co dimension. This causes the positive dot-waves to move in a linear direction at Co. Again the negative dot-waves mostly form a standing wave pattern with a net linear motion of Co with respect to the Co dimension. Thus the photon operates similar to a phase locked loop electrical circuit.

The property of mass is due to standing wave patterns between both the light speed Co dimension and the light speed Cs dimension. When we add photons to matter we get a combination of standing waves and moving waves. Yet photons add both linear motion and spherical standing wave patterns to particles.

The mathematics of the moving and standing wave patterns is beyond my engineering abilities. However I see what is happening and I can write simple algebraic equations to express the more complex wave equations.

As dot-waves are added to the photon, the wavelength will decrease and it will have a higher energy level. When dot-waves are radiated from the photon, it will have a lower energy level and the wavelength will increase.

As the photons travel from the sun toward the Earth they turns redder as the intensity of the Sun's gravitational field gets weaker. At the gravitational field center of gravity point between the Sun's gravitational field and the Earth's gravitational field, the photon's energy level is closer to red/white

Once the photons encounter the Earth's gravitational field alone the photons will turn toward the blue again. By the time the photons reach the surface of the Earth, the photons will appear white to us. That is the color of the spectrum of the light waves. Each individual photon's energy level has changed constantly during the trip. However the entire package looks white to us here.

The color of the light also changes when the Earth is moving toward the sun or away from the sun. When the Earth moves toward the sun, the combined gravitational field intensity increases and the photons turn bluer. When the Earth is moving away from the sun, the combined gravitational field intensity decreases and the photons turn toward the red.

Near the Earth, the Earth's gravitational field then becomes the reference point for the gravitational field intensity. Thus the color of the photons will be stabilized with very little variations as the photon moves from point to point on the Earth.

Einstein used the Michelson/Morley experiment to write his space-time equations for special relativity. He came up with independent reference platforms in his work. The math was a very good approximation to what was happening.

The speed of light Co is independent of any reference platform. It is a constant of the universe. The color of the light is variable. It depends upon the intensity of the gravitational field.

When the photon entered the instrument it was traveling at Co. It did not matter if the earth was moving toward the sun or away from the sun. The speed of light is referenced with respect to the dual light speed dimensions. Time is only a measurement and only a ruler and the speed of light determine the value of a second.

The work on Einstein and Lorentz was based upon a single light speed solution. Thus all the equations do not fully represent the working of the dual light speed universe.

If the distance was only traveled in the Co dimension, then Einstein's equations would be correct. For the dual light speed solution, Einstein's equations are a best fit approximation. If an object travels in the Co dimension some of the distance and the Cs dimension some of the distance, it would appear that the distance shrunk. Likewise if we look at the hands of a clock that is ticking in the Co dimension, the clock stops when it moves to the Cs dimension. Therefore when the distance shrinks, the clock slows. Thus the speed of clock ticks drop and Einstein says that time is elongated. Since Einstein's equations have excellent results, any other simple equations have to match his work.

In any event neither Einstein nor his distractors were absolutely correct. The two possibilities they argued over did not take into account and did not know that the photon jumps between light speed Co and light speed Cs. The only thing that changes is the photon's energy which depends upon the gravitational field intensity.

The experiment did not really test the nature of space and time properly. The problem is that the light waves constantly adjust to the gravitational fields that they travel on. Therefore arguing whether Einstein is correct or incorrect by means of an invalid experiment and invalid equations means nothing at all. In spite of this Einstein's work was a tremendous breakthrough in the understanding of the universe. His work is a best fit approximation for space and time.

Everything in the universe radiates Co/Cs dot-waves. This causes the universe to expand but it also causes the time clock and the ruler to expand as well. The light speeds Co and Cs cannot change. They define the universe we live in.

We see that the light from the far stars as gets redder the longer they travel. This is not caused by the common mode radiation of dot-waves from a Co/Cs configuration to the higher speed Cs/Co configuration. The photons continually lose dot-waves due to a second factor. As the universe expands, the gravitational field intensity drops.

As the photons breathe they continuously lose dot-waves due to the expansion of the universe at close to light speed Co. At big bang, the universe expanded at a very high light speed close to the geometric mean of Co and Cs. Over time the expansion stabilized at a speed near Co.

Today some astronomers see the universe expanding at slightly faster than Co. It may very well be that the expansion speed follows an exponential function with a limit at Co. Thus eventually it will stabilize at Co.

The Cs dimension of the universe is moving outward at Cs. Right after the big bang it was moving at the geometric mean of Co and Cs. Therefore the expansion of the Co/Cs universe is slowing while the expansion of the Cs/Co universe is moving faster.

A point will be reached where the Co/Cs universe will stop and the Cs/Co universe will also stop. This is when most of the Co/Cs dot-waves have become Cs/Co dot-waves. Then the universe will contract again.

Since the cycle time using a variable speed time clock is 1088 billion years, the universe will stop after 544 billion years. Of course this time when using our time clock is basically infinity. Thus it takes an infinite amount of time for the universe to cycle. Yet time is only a measurement so infinite time is meaningless. "It is what it is."

This expression tells us that the universe and existence are what they are. The universe exists because it exists. Man exists because man exists. There is no before time or after time. The universe always existed and will always exist. Man always existed and will always exist. The chicken and the egg always existed. Cause created effect and effect created cause. Everything is on a space time circle. Our lives are measured by our time clocks. We built the time clocks. A universe without man has no time clock. Likewise a tree that falls in a forest without man or animals makes no sound. Sound has no meaning without an ear to hear it.

As the universe expands and the gravitational intensity slowly drops, the photons lose dot-waves per unit time or unit distance. Particles radiate dot-waves and photons also radiate dot-waves as the material universe slowly erases as we head toward maximum size. During the voyage the photons will also gain energy as they near a star and lose energy as they move away from a star. This is the normal inhaling and exhaling of energy as the gravitational field intensity increases and decreases. Over the long term it is the expansion of the universe which slowly decreases the photon's energy level.

The dot-wave theory, and the conversion of mass to charge, shows that energy is volumetric in light speed. Thus:

(7-1) Energy = coulomb meters^3/seconds^3

Equation 7-1 tells us that energy is a charge that moves in three dimensions of space X, Y, and Z and three equivalent dimensions of time T_X, T_Y, and T_Z where the equivalent time is the three distances over light speed.

Energy does this by oscillating within the light speed Co dimension and then moving into the light speed Cs dimension and then oscillating within that dimension. Let us now look at momentum. Thus:

(7-2) Momentum MV = coulombs meters^2/ seconds^2.

Equation 7-2 tells us that the energy of momentum is limited to oscillations in two spatial dimensions such as X and Y while moving in the third dimension such as Z.

Let us look at the velocity in the X direction:

(7-3) Velocity V_X = meters/ second

Equation 7-3 tells us that velocity is limited to one particular spatial dimension/ time dimension.

From the understanding of the units, it is clear that energy and momentum are different geometric forms of the same thing. Energy can change into momentum and momentum can change into energy.

Energy which produces mass is caused by dot-waves which form standing wave patterns while momentum which produces photons are dot-wave patterns which move at the velocity C. When photons are added to an object we get combinations of standing wave patterns and moving wave patterns which increase both the mass and velocity of an object.

In classical physics we added momentum to an object and we noticed an increase in velocity. We then produced the laws of the conservation of momentum and everyone was happy. Then Einstein and others came along and we learned that mass and energy can be interchangeable. The atom bomb turned energy in the form of mass into pure photonic momentum. Spherical standing wave patterns of dot-waves were changed into planar patterns of dot-waves. Therefore energy can become momentum and momentum can become energy as the patterns of the waves change from standing waves to moving waves.

When we have an object moving with velocity V, the object has a mass M and a momentum MV. When we add photonic energy to it, some of the energy will go into the object in the form of heat. This will increase the mass of the object. Some of the photonic energy will go into the object in the form of momentum. The object now has two variables. Mass will increase and momentum will increase.

We now have a complex problem. We have a photon which converts into part photonic momentum and part energy. This tends to be a non-linear problem. There will be a Fourier series type solution to this problem. In Engineering School non-linear problems are solved by describing functions. We need a best fit solution to describe the non-linear addition of photonic energy to an object. The best solution would be a root mean square or geometric mean solution.

When the velocity is low, the complex problem does not exist to an appreciable extent as it is mostly a momentum problem. As the velocity increases to a moderate speed as a percentage of the speed of light such as the ground level of the Bohr hydrogen atom, we start to notice the increase in mass/energy of the electron as explained in Chapter 3.

At this time Einstein's equations start to work very well. Alternatively we can use the root mean square of the Doppler equations and get the same results as Einstein. Either solution is a best fit describing function solution to a more complex non-linear problem. Once we add wave analysis, the complexities require the most fantastic minds and computer analysis to solve. That is beyond my engineering mind's capability.

As we keep adding photonic energy to the object the mass keeps increasing but the simple Einsteinian relationship does not hold. Every photon added increases the momentum but does not keep increasing the mass after a certain point. We do not have to add an infinite amount of energy to bring the object such as a proton up to a maximum speed near light speed. In fact if we keep trying to add more and more photonic energy to the proton it will start to radiate any excessive photonic energy that we add.

At CERN they use powerful magnetic fields to insure that the photonic energy that is added stays with the proton. The net result is an increase of mass up to a maximum limit together with a huge amount of photonic energy. If we hit two high energy protons together we convert them into protons and anti-protons and a variety of interesting things.

The net result is that Einstein's equation for the increase of mass with added photonic energy only operates up to a limit. Beyond that limit the mass verses velocity curve levels off and no more mass can be produced. As we approach the speed of light we are in a very non-linear region and the atomic physicists can measure and observe interesting things.

We now need to know how much mass a photon produces when it attaches to an object moving at velocity V. If we found an asteroid that was moving at half the speed of light and build a laboratory to investigate the properties of space and time we would start out in a very distorted point on the curve. Einstein felt that all platforms were equivalent. That is only partially true.

The spherical patterns of the dot-waves get distorted when they are part of objects moving at high speeds relative to the center of gravity where they are. In fact no one could exist at such a speed. The Doppler distortions will not support human life at any speed beyond 0.2 C as

81

will be shown in Chapter 10, "Space Travel". Let us now look at the Postulates of Einstein.

Section 7-2 Einstein's Postulates

Postulate 1 specifies, that the Principle of Relativity stated that the laws of physics are the same or invariant in all inertial systems. The mathematical form of the laws remains the same.

This Postulate is not true in general. A platform moving at 0.9C has nothing in common with a platform moving at the speed of this Earth. The equations will be quite different. The question of how to define a platform is raised. Einstein produces platforms which are relative to each other. Thus his physics has no center for the universe.

His platforms are a way of producing describing function equations. They do not describe our actual Co/Cs universe. They are a particular approximation to areas of space that moves slowly in space compared to the speed of light Co.

We live on a spherical plane that is moving basically at the speed of light Co from the absolute center at big bang. Our gravitational fields are also moving at this speed. We do not consider this velocity to add to our motion in any of Einstein's equations. Thus as far as this outward velocity is concerned relativity is correct.

We exist in a dual light speed universe with gravitational fields. Every point in space has different gravitational field intensities. The net flow of the gravitational fields is like worm holes of sorts. These worm holes are the weighted vector sum of all the gravitational fields.

To find our reference point, we must find the weighted sum of the gravitational field's points of maximum dot-wave flow between dimensions. We must weigh the intensity each gravitational field by the inverse square of the distance from us. If we were near the center of our galaxy, that center would be the reference point. If we were near the sun, the sun would be the reference point.

Upon the Earth, the center of the Earth is the reference point. When we move toward the sun, the reference point becomes the gravitational balance point between the sun and the Earth. The reference point keeps changing.

When we want to know the velocity to be used in Einstein's describing function equations we must use velocities relative to the weighted center center of the gravitational fields nearest to us.

In general most objects near us are moving at very low velocity relative to the center of the Earth. Therefore for an area of the universe where things are moving rather slow as compared to the speed of light Co, Einstein's describing functions work quite well.

Postulate 1 tends to be true for similar galaxies and areas of space time where the velocities are low as compared to the speed of light. Einsteinian space-time is an excellent mathematical model for linear areas of space time such as where we live. Thus within the constraints of low speed platforms and linear areas of space time Einstein's work is a good approximation. The laws will fail for black holes and high velocity stars.

The laws of physics that we write down are best fit describing functions. If we attempted to write down all the laws and variations of the laws, then we could say that Postulate 1 would always be true. However this is not possible to do. Therefore Postulate 1 is a best fit solution for Einstein's work. It works well for linear space time. It is only partially true for the entire universe.

Postulate 2 by Einstein/Lorentz stated: The speed of light in a vacuum is a constant independent of the inertial system, the source, and the observer.

This postulate is true because whenever the dot-waves appear in the Co dimension, they travel at the speed of light Co. "They are what they are". Both Einstein and Lorentz were correct. As soon as we eliminate the time dimension we have no choice but to have a constant light speed Co dimension.

We need another dimension for the universe to work. Einstein chose the time dimension but that is not correct. It is the light speed dimension that is the correct dimension.

Since the light wave reformulates itself for each jump it makes between dimensions within the gravitational field the only thing that changes is its energy level. In addition the light from the sun flows along the suns gravitational field until it reaches our gravitational field. All along the way the photons keep losing and gaining energy levels as the photon changes from the light speed Co dimension to the light speed Cs dimension.

Thus the Postulate is reasonably correct for linear areas of space time although Einstein most likely did not understand that photons jump dimensions and gain and lose energy.

Lorentz was basically an electro-physicist which gives credibility to the postulate as well. In general electric theory is based upon a constant light speed and this works quite well.

When we look at our galaxy as a whole the galaxy is an independent inertial system with the center of our galaxy at a black hole. Thus our galaxy is a complete system and the only basic problem is its interaction with other galaxies. The light of the universe comes across to us but as we look at the picture of the universe we judge everything by our own galaxy and thus we stop at the point where we reach the speed of light Co. Then the astronomers believe they have reached the full extent of the light bubble but all we get is the limit of our own independent reference system.

Since the universe was expanding much faster than light speed Co/Cs at and after the big bang, we can never use our telescopes to see the big bang itself. If we build an indirect light speed Cs/Co instrument we may be able to get some additional data about the universe in the future.

The light speed Cs/Co photons have a linear wave moving at light speed Cs. When the wave crosses the boundary between the dimensions, it will travel at light speed Co. Then it will cross the barrier and return to the light speed Cs dimension where it will again travel at light speed Cs.

The light speed Cs/Co universe is mostly a photonic universe. It will not form particles such as we have. The masses are only at the tip of the waves like a tip of an arrow. We encounter these high velocity Cs/Co particle/waves and they affect us but understanding them is left to the scientists of the future.

The universe appears to us to be the distance Ru from the point of origin. However the picture of the universe shows that the actual distance around the universe is 2 pi Ru. That does not really change the calculations because what we see is all that we can see. This is because our gravitational field automatically changes the light waves from the other galaxies and references it to our galaxy. We cannot see the true picture of the universe. All we can see is what our eyes and instruments see and these instruments have been modified by the dot-waves as they jump between dimensions. Therefore another postulate is important.

Postulate 3: The universe that we see and measure is based upon our galaxy as an independent inertial system. As such we only see the part of the universe where the expansion reaches to the speed of light C. We cannot see beyond this point and thus the universe looks like part of a flat cylinder and not an ellipsoid. Thus we do not see where the ends of the universe curve around.

In order to compensate for our partial vision astronomers and scientists have imagined some sort of space time continuum whereas our shape is really spherical or ellipsoidal but we have no ability to see that.

Einstein got good answers but his theory was flawed. Yet he correctly saw that mass changes into momentum although he did not understand dual light speed energy. Einstein advanced us and confused us as well.

Let us now observe the effects of motion on an object. We know that in the lab, when an object moves faster with respect to the earth, the mass of the object increases as per Einstein's formula.

Let us move into pure free space when a photon is moving toward a stationary spaceship. Let us observe the motion of the photon and the spaceship from the Cs dimension.

In pure free space a photon coming toward a stationary spaceship would have a velocity of Co and the relative velocity would be:

(7-4) Relative velocity = Co

If the space ship was moving toward the photon, the relative velocity would be:

(7-5) relative velocity is Co+V

When we look at the relative velocity between the photon and the spaceship from the Co dimension we find that the relative velocity between the photon and the spaceship is Co. However when we look at the relative velocity from the Cs dimension we find that it is Co + V.

In order for this to be true, we find that the spaceship moved an additional time in the Cs dimension. From the Co dimension it looks like the distance has shrunk.

Since the spaceship is looking at the photon, the photon has turned toward the blue. As we head toward the sun, the sunlight turns toward the blue as well. This means that the moving gravitational field of the spaceship has grown stronger. Therefore the differential mass of the spaceship is larger in the front of the spaceship as the spaceship moves toward the stationary light source.

The spaceship now passes the photonic source. When the spaceship moves away from the photons, the relative velocity would be:

(7-6) relative velocity is Co-V

In this direction, the gravitational field intensity of the spaceship has been weakened or elongated. Therefore the relative speed as measured in the Cs dimension is C-V and the light looks redder with less energy. In addition the differential mass of the spaceship is smaller to the rear of the spaceship.

We then have a situation where the mass in the front direction of motion has increased and the mass in the rear direction of motion has decreased. We have a simple Doppler type problem. We also have Einstein's solution which produces a single mass increase with motion. Since his solution works quite well, the Doppler Solution must match Einstein's solution.

Einstein's theory of special relativity is a great approximation to the properties of space and time. Einstein's equations are a best fit approximation to the very complex non-linear solution. We can use his equations or the Doppler Equations.

First let us look at an equation for the conversion of a photon as it becomes part of an object. In this case it will become part mass/energy and part momentum. We will call the mass/energy gravitational mass/energy and the total combination of gravitational mass/energy and momentum we will call inertial mass. For this case inertial mass will be the amount of mass/energy if all the photonic momentum is converted into gravitational mass/energy which is spherical energy in the form of standing waves. For this analysis inertial mass will always be larger than gravitational mass.

It is important to note that a photon can be added to a hydrogen atom in a lower Bohr shell, and the energy of the photon will cause the electron to move into a higher shell. This energy forms a spherical rotational wave shape and this produces gravitational wave perturbations which increases the gravitational spherical energy. Thus if the electron has not changed the linear velocity of the atom appreciably, the photon has added mass energy to the atom.

In general the addition of a photon to a free electron causes an increase in the gravitational mass of the electron. At the same time, the photon converts some of its energy into the linear momentum of the electron. At the same time some of the angular momentum of the photon can cause the electron to rotate faster. Thus the photon can change the total linear momentum, the angular momentum, and the spherical momentum of the free electron. Thus the gravitational mass which is spherical momentum increases but not by 100 percent of the photonic energy.

Section 7-3: The conversion of photons into mass

Photons are planar momentum sources. They are dual positive/negative dot-waves which move at the speed of light Co while the oscillation into the Cs dimension is phase locked. When the positive dot-wave structure is moving forward in the Co dimension, the negative dot-wave structure is in the Cs dimension and moving with it. Then the negative dot-wave structure continues on the forward path while the positive dot-wave structure is locked to it in the Cs dimension.

When a photon is added to a mass, the velocity of the mass will increase and the velocity of the photon will decrease rapidly. The photon has planar energy which consists of linear and angular energy. This is moving wave energy. The photon does not contain any spherical standing wave energy. Thus a photon has two-dimensional energy patterns whereas a particle has three-dimensional energy patterns.

As the photon is added to a mass, the velocity of the photon slows and it becomes part gravitational mass and part angular momentum and part linear momentum. The linear momentum is no longer in the form of a photon traveling at light speed Co. Now it exists as a part of an object traveling at velocity V along with parts of many other photons.

The photon acts like a photon motor but it is now constrained to a low speed. As far as the part of the photon is concerned its axial velocity has been reduced to the modified speed of V rather than Co.

If we keep adding photons to an object its linear velocity will increase. At the same time its mass will increases as well and the net angular momentum will also change. If the object goes fast enough it will start to spin rapidly. The increase in gravitational mass can be caused by the conversion of photons into spherical energy directly. It can also be caused by the speed of electrons within their Bohr orbits moving closer to the protons. This causes greater spherical energy to occur. Thus the forcing of additional linear energy into a fast moving object will tend to bring all the electrons down past their lowest state and thus increase the mass of the electrons

As we force more and more energy into the object a point is reached where we attempt to bring the atoms into a state of temporary neutrons. Thus we are crushing the energy into a smaller package. This is only a temporary state and a point is reached where the object will readily radiate any additional energy. We are also producing a situation where the object is ready to explode.

The first photon will have the greatest impact on the increase of mass. The second will have less effect since the velocity has risen and more of the second photon remains as linear and angular energy.

As more and more photons are added, the object say a proton is moving closer to light speed. Each additional photon raises the total momentum but only raises the gravitational mass a little bit. Thus more and more energy stays in the form of both linear and angular momentum and less as spherical momentum.

When we look at the cyclotron at CERN we find protons spinning around at basically light speed. At this speed the protons will radiate as many photons as they absorb. The very strong magnetic field manages to keep the photons in line with the proton.

Einstein's equation fails as we reach close to light speed. In fact it fails somewhat before light speed. His equation assumes that you could keep adding photons to a proton and the mass of the proton will keep increasing. The reality is different.

At very low velocities classical physics explains the increase in momentum with velocity. The mass increase is basically insignificant. In the Bohr hydrogen atom, the Einsteinian mass increase in the electron is evident at $C/137.036$

The problem we have is that if we keep adding huge amounts of momentum into the proton, the additional mass increase will not match the additional velocity. Instead of the photons producing stronger spherical standing dot-wave patterns within the proton, they remain mostly photons and radiate away readily.

In any event Classical physics works up to a certain velocity. Then Einstein's laws work well up to a higher velocity. Finally we end up with the inability to create any more gravitational type mass and the additional momentum remains as momentum.

What is happening at CERN? The scientists are pumping up a proton to a very higher mass/momentum level. Then they keep adding photons and the mass no longer increases but the momentum only increases slightly. When this huge mass/momentum combination hits another proton which is moving in the opposite direction matter and anti-matter and a lot of interesting things are produced. The result is that high mass and even higher photonic energy created matter and anti-matter and other unstable particles.

Matter and anti-matter are like electrons and positrons. Thus photons produce particles and antiparticles. Waves moving at the velocity Co produce standing waves that are particles and antiparticles.

Scientists would hope that we could go back in time to the original protons and neutrons. However the energy levels of these things were much higher than we can produce now. The universe is larger now. The ruler is larger now and the spacing of the gravitational fields is larger now. So we cannot produce a spherical gravitational field that will permit an ancient proton to survive for very long. All we can produce is the partial mass of an earlier proton which is combined with huge amounts of linear and angular energy. Therefore this high energy proton is similar but quite different than one billions of years ago.

Now let us look at the Doppler mass equations which match Einstein's Equations. Both Einstein's Equations and the Doppler Equations produce the same describing function for gravitational mass increase with velocity.

Section 7-4: Doppler Mass

The requirements for the Doppler mass are similar to a moving radio source. In addition the geometric mean of the frontal mass and the rearward mass must equal Einstein's equations. Einstein has a best fit describing function so the effect of eliminating the time dimension and using the dual light speed dimensions must produce the same results.

In the future mathematicians can produce adequate equations to do an equal or better job but it is very difficult to produce these equations at this time.

In Doppler Space Time the change of mass with velocity follows the same rules as with moving radio sources. In general for a constant light speed universe, the speed of light in the Co dimension is constant. The measured speed of light in a vacuum only depends upon the speed of light of space itself.

When we look inside of empty space, we find dot-waves which are oscillating from light speed Co to light speed Cs. The fundamental dot-wave is as large as the universe itself. The frequency of vibration is extremely low. The gravitational fields have very low fundamental values. The same is true of the positive electric field and the negative electric field. There are all vibrating as well between the Co dimension and the Cs dimension.

This also means that there are dot-wave current flows. A point source has a current flow. Therefore there are both electric field point sources and current sources as well. People look at the charge Q standing in space and cannot readily see the current flow out of it. Yet for the dual light speed universe, the charge Q has a very large current flow through the worm hole as the time is very fast and the charge at the point in the Co dimension decays very rapidly.

For the single light speed solution, it looks impossible that a charge Q would have a current flow. Yet for the dual light speed solution, current flows out of a charge Q constantly. It flows from light speed Co dimension to the light speed Cs dimension and back continuously. As such we can say that the charge Q is really a current source. This has to be included in future electrical theory courses.

The vacuum of space is quite alive. The whole universe is composed of dot-waves. Two basic things make everything. There is a lot of dark matter in space. This is basically the Co/Cs dot-waves which make up the fields. There is also a lot of dark energy as there are many particles and photons of Co/Cs energy. In addition there are photons of Cs/Co energy. These tend to be photons although they carry with them a small degree of mass. At the transition point at the beginning and end of the Cs/Co wave. Empty space is not empty at all.

Once you fill space with a lot of heavy particles, the speed of light drops. In water the speed of light is less. In glass it is less as well. The speed will be less in air and will also vary with the temperature of air. Thus Einstein specifies that the measurements should be taken in vacuum.

If you look into a black hole, the super density of the gravitational field causes the light to slow and stop. The black hole has a very powerful worm hole as energy flows between the light speed Co dimension and the light speed Cs dimension. This energy moves back and forth but the radiation from the light speed Co dimension to the light speed Cs dimension is very high. The net result is that the black hole eats up the light speed Co/Cs universe and causes light speed Cs/Co photons to fill the light speed Cs dimension.

Some scientists believe that the worm hole inside the black hole will enable people to travel to a different area of our universe. This is merely fantasy. Anything entering the worm hole is reduced to light speed Cs/Co photons and death. It is no different than putting yourself in a large blender and turning on the power. You end up rapidly as little bits and pieces of yourself.

The measured speed of light is Co = 2.99792E8 meters per second. This speed is very slightly lower than the maximum light speed found in areas of space with minimal amounts of particles per cubic meter. Yet this is the best that we can readily measure.

The reason I look for an ideal light speed is that when we use numerical analysis on the constants of the universe, the numbers work out better when the light speed is slightly larger then used at present. Thus my several years of numerical analysis starting in 1981 provided me with various relationships which were more perfect when the light speed was very slightly higher than measured here. Unfortunately I no longer have my notebooks from 1981-3 because I have very limited space for my books and notebooks.

Let us look at a mass moving toward the Earth with velocity V. We can look at the mass as a type of equivalent frequency source. The incoming equivalent frequency of the mass is:

(7-7): $f = f_o (C/C-V)$

If the mass was moving away from the Earth the frequency would be:

(7-8): $f = f_o (C/C+V)$

The root mean square of the signal would be

(7-9): $f_{RMS} = f_o / [1-(V/C)^2]^{0.5}$

These equations show that a frequency source moving toward you has a higher frequency than a frequency source moving away from you. The root mean square frequency heads toward infinity as the velocity approaches the speed of light. If we look at a mass and keep adding photons to it, the photons slow and turn into partial mass partial photons. This will increase the mass of the object. The Doppler mass equations are:

(7-10) $M_{FRONT} = Mo\ C/(C-V)$

(7-11) $M_{REAR} = Mo\ (C/(C+V)$

(7-12) $M_{RMS} = Mo / [1-(V/C)^2]^{0.5}$

Equation 7-10 specifies that the forward mass will raise toward infinity as the velocity approaches the speed of light. Equation 7-11 specifies that the rearward mass will drop to 0.5Mo as the object approaches light speed. Finally Equation 7-12 specifies that an object will have a root mean square mass or geometric mean mass which rises toward infinity as the velocity reaches light speed.

Equation 7-12 gives us the geometric mean mass. What does this mean? It is the same formula that Einstein found as he explained his special relativity. Equation 7-10 and Equation 7-11 are the most simple Doppler type equations whose root mean square is identical with Einstein's equation.

To me the Doppler Equations represent what is actually happening. Since Einstein's equation works for the hydrogen atom we can only conclude that the geometric mean mass of equation 7-12 is the gravitational mass. Thus:

(7-13) $Mg = M_{RMS} = Mo \: / \: [1 - (V/C)^2]^{0.5}$

Experiments have shown no difference between the gravitational mass and the inertial mass. The only problem we have is the combination of standing wave energy and moving wave energy. In general, physicists compare the total energy of the two forms of energy. Then we can calculate an equivalent mass for the combination of the two energies. Yet at this point these equations are left to the mathematicians to solve for the dual light speed universe.

Section 7-5 Doppler Center of Gravity

Another problem with Einstein's work is the failure to understand the Doppler shift in the center of Gravity of a moving object with velocity V. As an object moves its forward gravitational mass increases while its rearward mass decreases. The faster an object moves the greater the forward mass. Therefore speed is accompanied with a shift in the center of gravity of an object. For a relatively low velocity, the shift can be approximately calculated as follows.

(7-14) M_F = Mo $[1 + (V/C)]$

(7-15) M_R = Mo $[1 - (V/C)]$

If we assume that the length of the object is L and that the front mass is at a distance $L/4$ from the center and the rear mass is at a distance $L/4$ from the center, then the shift in the center of gravity would be:

(7-16) Center of Mass shift = $+L(V/4C)$

The importance of this is that a moving object has a Doppler shift of the center of mass of an object. This tends to keep an object moving at a constant speed. If the velocity gets very high the above equations have to be replaced. In addition the length of an object also has Doppler effects so the entire problem gets more complex. Now let us look at the length of an object as it moves toward the speed of light.

Section 7-6: The Doppler Length

Einstein's formula for the size of a gravitational mass is:

(7-17) L_{Xg} = Lo $[1-(V/C)^2]^{0.5}$

This formula states that when we add energy to an object, its mass increases and its size shrinks. If we achieve light speed then the gravitational size will be zero. We now want to know what the Doppler length will be. We know that the Doppler mass in the front is larger and therefore the corresponding Doppler length in the front must be smaller. The geometric mean of the frontal length and the rearward length must equal Einstein's equation. Therefore the best fit equations are:

(7-18) Doppler Length Front = Lo [C-V]/C

(7-19) Doppler Length Rear = Lo [C+V]/C

(7-20) Lg = Lo / [1 – (V/C)^2]$^{0.5}$

For this solution the frontal Doppler length approaches zero as the velocity approaches the speed of light Co. At the same time the rear Doppler length goes toward twice the rest size as the velocity goes toward light speed Co. In effect a proton moving in the cyclotron gains a tail.

We can now make a chart of the various lengths verses the velocity.

Figure 7-1 Gravitational and Doppler Lengths verses Velocity

V/C	L$_{GRAV.}$	L$_{FRONT}$	L$_{REAR}$
0	1.000	1.000	1.000
0.1	0.994987	0.9	1.1
0.2	0.979796	0.8	1.2
0.5	0.866025	0.5	1.5
0.9	0.435890	0.1	1.9
0.99	0.141067	0.01	1.99
0.999	0.044710	0.001	1.999
0.9999	0.014142	0.0001	1.9999
0.99999	0.004472	0.00001	1.99999

It is interesting that the gravitational length shrinks toward zero as the velocity approaches the speed of light Co and at the same time the rearward length reaches toward twice the original size.

What does this mean? A proton in a cyclotron moving near the speed of light C has almost no size when it approaches a receiver. Yet when it passes the receiver by, it looks like twice its normal size.

We can now write the equations for the time clock.

Section 7-7 Einsteinian Time

Einstein's formula for the time clock is:

(7-21) $Tg = To/ [1 - (V/C)^2]^{0.5}$

Einstein's formula states that the time between the ticks of a clock within a moving object will take longer to occur as the velocity of the object rises toward the speed of light Co. Thus the clock slows. This information was verified by experiments with satellite clocks. Notice that when you reach Co the clock stops. Einstein stated that if you could exceed light speed you would go back in time.

To Einstein time was a dimension. Yet that is false. We only have two light speed dimensions. The light speed Co is only the lower light speed.

To say that yesterday exists is meaningless. It is only the memory of yesterday that exists in photonic information. You cannot reverse time because time does not exist. You cannot physically exceed the speed of light in the Co dimension.

We can produce Doppler forward and rearward lengths but we cannot produce Doppler forward and rearward times.

If we look at the moving clock, as the velocity increases we spend a greater percentage of the trip in the Cs dimension. When we are in this dimension, the clock in the Co dimension does not move. Therefore the clock ticks slow down because the clock stops for a greater percentage of the journey.

In Einstein's space time he felt that if you could travel faster then light speed Co, you could return to the Earth at a prior time in history.

In order to travel near the speed of light, the Doppler distortions will cause human life to perish and the spaceship to blow up. Protons and electrons can handle the differences in frontal mass and rearward mass and frontal length and rearward length. Spaceships cannot hold together with these great Doppler mass and length differences.

In spite of that, what defines an event? A star explodes and light speed Co/Cs photons travel outward at the speed of light Co. At the same time light speed Cs/Co photons travel outward at the speed of light Cs. In the Cs dimension, the information concerning the explosion travels 18832.8 times as fast as in the Co dimension.

Even if you could approach the speed of light Co, you can never reach the event before the event occurred. To do that you would have to travel faster than light speed Cs. Yesterday does not exist. The only thing that exists is the memory of yesterday in photonic energy.

The star that exploded a light year away a light year ago is reaching us now. Soon the information will be gone. We could record it on memory devices. A year from now we could go back to our memory devices and see the explosion.

It is true that when a person goes on a spaceship that moves at a very high velocity that the time clocks will slow. At the same time the bodily time clocks will slow as well.

On the surface it may appear that the person could live a little longer due to the increased speed. However the Doppler variations may cause bodily damage over long periods of time. The clock may not care that it slowed down but our bodies may suffer prolonged periods of distortions.

Einstein also felt that the maximum relative velocity is limited to the speed of light Co. Two objects approaching each other at Co each would only have a relative velocity of Co.

That may be good for a single light speed universe. It does not work for a dual light speed universe. When we look at the two objects from the light speed Cs dimension, it is clear that they are traveling toward each other at 2Co.

When the objects are very close to each other, ordinary linear space time will not work quite so well. There will be distortions in the Co dimension gravitational fields of the two objects as they approach one another. This is also true in the Cs dimension gravitational fields.

We then have a complex non-linear space time problem. This will aid and assist the destruction of the two objects as they get close to each other.

Section 7-7: The Motion of Objects in a Straight Line with Constant Velocity

In classical physics, an object moves in a straight line and at a constant velocity unless acted upon by a force. The question is why is this so?

Doppler Space causes this to be true. When the front Doppler mass is larger than the rear Doppler mass, the object has an unbalanced momentum which tends to keep it at constant velocity unless the Doppler differential momentum can be reduced by a force. This is especially true when we realize that the image of the object in the Cs dimension has the same differences although the masses are much smaller.

When we look at the frontal Doppler length, we see that it is smaller in the front and larger in the back. These differences are even greater in the Cs dimension. As we approach the speed of light Co, the frontal image is basically zero but the rearward image is twice the original size. The object then is part of an image that is twice as large as itself. This causes it to move in a straight line. When we add the Doppler length in the Cs dimension we find that the object has a much longer tail.

For the case of the photon, the image in the front will be zero. When we go to the rear, the photon has a length of one wavelength and an image of two wavelengths. This by itself would assure that the photon would move in a straight line. However once we add the light speed Cs image to the photon, the total length of the photon will appear to be 18832.8 wavelengths for the ordinary image and twice as much for the Doppler image. Thus the photon will stay in a straight line due to its very long tail.

When you add photons to an object, the Doppler differential mass increases and the Doppler image gets larger as well. The momentum increase is stored in both the differential mass increase and the length increase.

Einstein failed to understand the Doppler masses and the Doppler lengths. Much of his work was excellent because his mathematics was the geometric mean of the Doppler. The simple Doppler equations shown in this chapter will be corrected in the future but they suffice for this Engineering analysis. They show what is happening and how the universe works. Once the mathematicians and physicists understand how the universe works, they can greatly improve the understanding of the dual light speed universe for themselves.

The problem with that is that it limits the understanding to a select few individuals who can contemplate such things. This book is written at a level where the average bright high school student can understand how the universe works.

Chapter 8: Angular Doppler Space Time

Section 8-1 Introduction

In many respects Angular Doppler Space Time is merely a variation of Linear Doppler Space Time. As the Earth travels around the Sun it has an Einsteinian gravitational mass increase due to its velocity. It also has a larger Doppler mass in front of the motion and a smaller Doppler mass to the rear of the motion. The differences in the Doppler masses insure that the Earth will continue in its orbit and not fall into the sun.

The Sun is continuously pulling on the Earth due to its gravitational field. In reality the radiated gravitational dot-waves from both the sun and the Earth push the Earth toward the Sun and vice versa. The Earth is also spinning on its axis. Every atom upon the Earth experiences its own gravitational mass increase due to the rotation of the Earth. At the same time every atom also experiences differential Doppler masses. Therefore at any instant of time in the rotation, the increased mass and size could be viewed as a linear space time problem.

For the linear problem we know that the object is shrinking gravitationally. Therefore a rotating object will have a shrinking radius as it moves faster. Since the outer radius is moving faster, it will tend to shrink more. The inner radius will tend to move slower and therefore it will shrink less. We then have a problem in which as an object speeds in rotation, the outer surface is crushing itself against the inner structure. The net result is that crack lines will form and the object will explode at a certain rotational speed.

A mass moving in a linear direction tends to hold together at low velocities fairly uniformly. As the velocity increases the compression at the front and the elongation at the rear tend to rip the object apart at very high velocities.

Rotating objects have much more severe forces acting upon them. If you have a stone grinder flywheel, it will burst at a certain rotational velocity which has a certain surface speed. If you take that flywheel and put it on a rocket at a huge linear speed, the flywheel will not be damaged until you go to very high speeds. Therefore objects are more stressed with surface velocities for angular space time than with linear space time.

Let us look an object that has a thin shell such as a tube that is rotating. As the tube rotates faster and faster, the atoms will shrink gravitationally. We can use the formula for length in Chapter 7 and change L to R. The gravitational radius will be:

(8-1) $Rg = Ro \, [1-(V/C)^2]^{0.5}$

In equation 8-1 we see that as the surface velocity increases. The radius of the tube shrinks. Since the surface velocity is the radius times the radian frequency, we could use the rotation in the equation.

(8-2) $V = R \, w$

In Equation 8-2 R is the radius and w is the radian frequency. However for this Chapter I will use the surface velocity to describe the angular motion. The rotating object will have both a gravitational mass and Doppler masses just like an ordinary mass with linear velocity.

The space time equations in this Chapter do not apply to the shape of slow rotating objects such as the Earth. The Earth is elongated in diameter at the equator and flattened at the poles. This is all due to classical physics. It will obey the above space time laws but that is only a small part of the physical equations. For this Chapter only objects with strong bonds such as a steel pipe will exhibit radius space time shrinkage with velocity. In addition the protons and electrons themselves are subject to the space time equations. The same is true of the entire universe. It should be noted that I call the shrinkage space time shrinkage since it has a common understanding even though time is only a measurement. In the future some other word will have to be used.

The Doppler radius at the center of rotation is:

(8-3) Inner Doppler radius = Rg [(C-V)/C]

(8-4) Outer Doppler radius = Rg [(C+V)/C]

When an object rotates its gravitational radius Rg shrinks. At the same time its inner Doppler radius shrinks even more but its outer radius expands. At near the speed of light its inner radius will be zero but its outer radius will reach toward twice the rest radius.

In the same regard a rotating object at high speed will have a similar mass increase to that of an object moving in a linear fashion.

We see that rotational energy is stored in the Doppler fields. If we look at a gyroscope that is spinning rapidly it ability to maintain its level position is due to the Doppler plane that is produced as it spins. This Doppler plane of photonic energy will struggle to prevent any axial Z axis rotation when it is spinning in the X/Y plane.

The rotating tube has additional destructive forces acting upon it from the Cs dimension. As the tube spins faster and faster, it will shrink gravitationally. As the speed gets very high we will get additional destructive forces in the Cs dimension.

When we look at the Bohr orbit, the stability of the Bohr orbit is caused by the light seed Cs dimension where the proton is rather large as compared to its tiny size in the

Co dimension. Similar effects produce additional repulsive forces in the Cs dimension of the spinning tube. This will cause the tube to start to be destroyed by Cs forces before it starts to be destroyed by Co forces.

Section 8-2: The Gyroscope

The gyroscope spins and acts to maintain it spin velocity and it's the stability of the plane of rotation. The Doppler radius acts to make the image of the plane of the gyroscope larger than its rest size. At near the speed of light the Doppler radius would be twice the size. When we look at the image in the light speed Cs dimension, the image of the gyroscope radius and the associated gravitational fields is 18832.8 times as much for the stationary image and twice as much for the image when the speed approaches light speed Co. Of course it would not survive such speeds before breaking apart.

In any event the stability of the gyroscope is due to its Doppler image and its interaction between the light speed Co dimension and the light speed Cs dimension.

It should be noted that most of the action in the light speed Cs dimension occurs at the atomic level and atomic distances. The gyroscope has all these effects occurring at the level of the radius of the proton and the Bohr orbit.

When we look at the size of a spaceship, all the effects occur at small distances but the gravitational fields extend to large distances. The net result is that when we say that a length is very much longer in the Cs dimension, it is really the interactions with the Dot-waves in space which are so much longer.

Chapter 9: Spherical Doppler Space Time

Section 9-1: Introduction

The dot-waves and ordinary particles illustrate the properties of spherical Doppler space time. At the big bang, the light speed Cs/Co dot-waves were compressing toward a common pinpoint at the junction between the dimensions.

The big bang inverted the Cs/Co dot waves through a worm hole between the dimensions. The big bang looks like an explosion to us but it was a very high light speed flow of energy. As the energy passed through the worm hole, it was inverted.

Everything did not get inverted. There were a small percentage of Cs/Co dot-waves that remained. They passed through the worm hole and this expanded the universe at a very high inflationary rate.

In many respects each particle such as an electron or a quark looks like a mini big bang as far as the inversion is concerned. Light speed Cs/Co energy in the particle passes through the mini worm hole at the center of the particle. However the particle does not explode. It oscillates and inverts over and over again through its mini worm hole.

The big bang works in a similar manner but it explodes. Prior to the big bang, the Cs/Co dot-waves existed over a large volume of space. This volume kept getting smaller and smaller as it moved at light speed Cs.

Each dot-wave had its own worm hole. As the universe compressed toward a pinpoint, many worm holes became a single worm hole. Through this singular worm hole, the Cs/Co energy flowed. This became Co/Cs energy in the Co dimension. A singular black hole was forming. It was oscillating at a high rate.

More and more light speed Cs/Co energy flowed through the common worm hole. The frequency of oscillation increased. The mass of the universe at the common worm hole grew larger and larger. The temperature rose to a very high level.

The net result was a very small neutral particle. A point was reached where it exploded. At the same time additional Cs/Co energy was flowing through the worm hole. This caused the universe to expand rapidly at a large expansion speed. Thus:

(9-1) Expansion speed = $(Co\ Cs)^{0.5}$

(9-2) Expansion speed = $137.23Co$

The most likely solution for the expansion speed at the big bang is the geometric mean between light speed Co and light speed Cs. Since Co = 2.99792E8

(9-3) Cs = 18832.8Co = 5.64592E12

And

(9-4) Expansion speed = 4.11405E10

The inversion and oscillation at big bang most likely turned into spherical planes. We had a light speed Cs/Co plane which turned into a Co/Cs plane when it inverted. The final explosion most likely occurred upon the Co/Cs spherical plane. A singular oscillating black hole became billions of black holes centered upon the spherical surface. We then got billions of black holes that exploded to form the billions of galaxies that we have.

The billions of early black holes which were the center of the forming galaxies exploded leaving behind billions of lower energy black holes within each galaxy. In addition thousands of billions of lower energy mini-black holes formed at the centers of forming stars.

Within particles the dot-waves form spherical surfaces that oscillate from a maximum radius in the light speed Co dimension to a pinpoint. They pass through a worm hole and invert as they enter the light speed Cs dimension. The particles will not explode unless additional spherical energy is put into them. This requires spherical electromagnetic fields and spherical laser photonic patterns. Someday this will be accomplished and we will obtain the energy from the proton as it self-destructs.

The equations of spherical space time are similar to the equations of angular space time.

(9-5) $Rg = Ro \, [1-(V/C)^2]^{0.5}$

In equation 9-5 we have a mass that is oscillating from a radius Ro toward the center point at radius zero with a velocity V. The gravitational or Einsteinian radius is shrunk due to its velocity. The corresponding Doppler radiuses in the direction toward and away from the center point are:

(9-6) Inner Doppler radius = $Ro \, [(C-V)/C)]$

(9-7) Outer Doppler radius = $Rg \, [(C+V)/(C)]$

The Doppler forward radius in the direction of the center shrinks as the velocity V increases. At the same time, the outer radius expands. The geometric mean shown in equation 9-5 shrinks with velocity.

As the velocity of a particle approaches the speed of light Co, the inner radius heads toward zero and the outer radius will head toward twice the rest radius.

When we look at the particle in the Cs dimension we get similar equations. Somehow the mass of the electron and

110

the various quarks is related to the light speeds Co and Cs and the Doppler images of them in the Co dimension and the Cs dimension. This requires further study of the dual light speed universe.

When we look at the atomic bomb, spherical compressive forces were exploded forcing the mass to be contained in a smaller space. This caused the inner Doppler radius to be reduced and the outer Doppler radius to be enlarged.

These atoms were ripped apart due to the Doppler mass differential and the Doppler radius differential. Once we add the Cs dimension into the mix, this increased the unbalance.

Chapter 10: Space Travel

Section 10-1 Introduction

One of the hopes and dreams of man is to be able to travel the stars in search of new planets to inhabit. As we look around our area of the universe we do not find any viable new Earths in our vicinity. We do find possibilities of placing colonies on nearby planets and moons.

It is difficult for us to travel to Mars. Yet we can do it someday. The biggest problem is the necessity for an engine that can use atomic energy for propulsion.

The energy of the protons and various sub-particles was locked into them after the big bang inversion during subsequent energy implosions. Most of this energy was locked into the protons during the explosion of forming galaxies.

In order for us to travel to nearby planets and build colonies, it is necessary to develop the proton thruster rocket engine. The black hole breaks apart some of the protons and sends much of this energy through the worm holes into the Cs dimension. At the same time it produces huge amounts of ordinary photonic Co/CS energy which radiates into the Co/Cs dimension. Thus we find black hole radiation in our primary dimension. Some energy remains trapped in the black hole and in turn this energy will radiate away. The black hole slowly eats the Co/Cs universe and feeds the Cs/Co universe.

The proton thruster engine would provide the energy to achieve a trip to Mars in a reasonable time. In order to produce the proton thruster engine it is necessary to develop a mini-black hole engine which will break apart the protons and radiate light speed Co/Cs photons and light speed Cs/Co photons as well.

112

We need to achieve velocities of nearly 2 million miles an hour in order to achieve Mars in a few days. We also want to build spaceships capable of reaching 0.2C in order to study the effects of velocity on the human body and our robots.

In order to aid this process, it will be necessary to produce light speed Cs/Co telescopes. These instruments will pick up photonic emissions at light speed Cs and convert them into readable Co/Cs images. Radio and television transmissions exist in the Co dimension and the Cs dimension. Future light speed Cs receivers can tell us if any planets with man or higher life exist by their radio and television transmissions.

In addition when our spaceship travels at 0.2Co, it is important to spot any masses in the path of our spaceships. Cs/Co radars will be able to pick up any problems before the space ship arrives there.

Cs/Co laser beams will be able to destroy space rocks and other damaging masses well before the spaceship reaches the area. Cs/Co technology is for future man upon this Earth. Unfortunately this technology can produce much more powerful military weapons than exist now. These weapons can readily destroy incoming missiles as the photonic rays will destroy all the binding energies of the incoming rockets. On the positive side they can save lives. On the negative side they can destroy cities and countries. Thus the future Cs/Co technology is a two edge sword. One side is for good and the other side is for evil.

It may be possible that higher light speed planets could exist with superior life forms. These planets would be mostly photonic. Since the mass is very low, they are mostly energy planets. The life forms would tend to have superior existence. Their minds would tend to be collective minds. It would be one planet and one total mind. The life forms would all work for the collective good. Someday when we have better instruments we might be able to communicate with these planets. It may also be possible that we could achieve these planets. For that to occur, our stored intelligence would have to be transmitted to these planets and placed in a new photonic body.

We would call these higher intelligent life forms aliens. They could arrive on planet Earth aboard their light speed Cs/Co spaceships. We would not readily see these spaceships. However we will interact with them slightly. They could grab a person and copy his mind. We have a light speed Co/Cs mind. It is actually two minds within one dual light speed body.

The aliens can read the data within the light speed Cs dimension of our mind. Then they can put the data in storage. When they return to their planet, they could put the data in a new body. Thus they can copy the mind of anyone they choose. In our sci-fi movies we contemplated the body snatchers. Yet the aliens do not want our bodies. They want our minds and they will copy them. We live at most a hundred years but the aliens have superior bodies and can live a thousand years or more.

Many people claim they had alien encounters. Some have ended up on a spaceship. They were studied and let go. As they were studied, their minds were copied. The people returned to this Earth but a mental copy of them was brought to the alien planet.

An alien can enter our bodies and take over our minds if it so chooses. It only involves inserting and removing data from our minds. We then become a mental alien. Since they are superior creatures of one collective mind, they tend to be kind and do well. Yet it is possible that a planet with evil aliens exist. They could take over our mind and create hatred and wars.

In any event the physics of the dual light speed universe permit such things to happen. In order to detect aliens we would have to build superior instruments which look at the light speed Cs/Co universe upon this Earth. This will detect aliens and alien spacecraft. Right now we have no positive proof of such aliens.

Section 10-2: The Proton Thruster Engine Requirements

The proton thruster engine is the most important physical device for the trip. We must be able to travel up to 20 percent of light speed (0.2C). In addition we must maintain continuous acceleration at 32 feet per second per second (1G). Weightlessness will not be acceptable except for short periods of time. Right now how long a person can stay in space in a weightless condition is being studied. So far we have achieved a year. Yet this is not desirable for the majority of people who want to go to Mars. This problem will be eliminated if we always keep a spaceship under continuous acceleration.

In order to accomplish 0.2C, we must accelerate uniformly at our rate of gravity of 32 feet per second per second. This will take approximately 2.5 months. Then we must slowly turn around and slow for 2.5 months in the reverse direction to return to the Earth.

The ship will have two main proton thruster engines. The rear one will bring us up to 20 percent of light speed. The front one will protect us against anything in our path.

As we travel near 20 percent of light speed, we will produce magnetic and photonic waves in both the front and the rear of us in the light speed Co dimension and the light speed Cs dimension. These will act as protective barriers as well. The front of the spaceship must come to a point, which provides a degree of protection when used in combination with the protective waves we produce. If we find a rock ahead of us, we must increase the energy of the front atomic engine and the rear atomic engine in synchronization. The total power will increase but the crew will still experience a force of only one G. The front proton thruster engine will be focused into a ray that completely destroys the rock and everything will be safe.

Of course we cannot destroy a huge space rock. Therefore we must always look ahead and see what is there. If a space rock is reflecting sunlight, then we can visibly see it. If a dark huge rock is in our path, we will not see it. However high light speed radars will send out waves that will rebound from it and our computers will have to decide what evasive action is necessary.

The engine will permit high-speed space travel up to 0.2C without the necessity of weightlessness. Ordinary people can ride the spacecraft. Ordinary people can travel to Mars. Takeoffs and landings are easy. We can ride the beam up, and ride the beam down when we land.

The atomic engine permits easy space travel. You can leave the Earth at 1.5 G's or less. A man who weighs160 pounds on the Earth will weigh 240 pounds during takeoff.

The proton thruster engine will require an oscillating spherical electromagnetic field together with spherical laser fields surrounding the fuel. The spherical forward Doppler mass of the protons within the fields will bring the protons up to the point of destruction.

This will cause the Co/Cs proton structure to break apart along with all the sub-particles. The net result will be light speed Cs/Co photons as well as light speed Co/Cs photons. At the same time positrons will be produced which will attract surrounding electrons to produce more Co/Cs photons.

The Engineering details of the engine will require teams of research Engineers and scientists to produce the correct design. It will be a long term effort and a lot of experimental work has to be done.

In the end, the world will have abundant energy from the proton but at the same time powerful weapons of mass destruction can be produced. So the technology is a two edge sword. On one side is abundant cheap energy and on the other side is destruction.

Section 10-3: Trip to the Moon

The trip to the moon tests out the spacecraft design. It gives us the opportunity to take off from the Earth and to land right back upon the Earth. We will come down backward riding the beam downward while gravity pulls us down. The front beam will also be used to prevent the spacecraft from tilting.

The spacecraft can land on any hard rocky surface. It could also melt various surfaces and produce its own landing floor. For the Earth one or more sites will be used and precision corrections must be made during the spacecraft flight.

The time to reach the moon and the peak velocity is found by the simple physics formulas for constant acceleration. They are:

(10-1): $S = \frac{1}{2} At^2$

(10-2): $V = At$

The moon is approximately 240,000 miles away. Half the distance is 120,000. The time to achieve half the distance is found by equation 10-1. This amounts to 6290 seconds when the acceleration is 32ft/sec^2. It takes 1.75 hours to reach half the distance to the moon. The speed is found from equation 10-2 to be 137,400 miles per hour.

At the halfway point, the front engine must produce a deceleration of 1G. The rear engine will be reduced to a small value to maintain stability. It will take another 1.75 hours to reach the moon. The entire trip will take 3.5 hours in comfort. Of course it will take time to turn around and land. At least an additional hour is required during landing.

The only discomfort will occur during the changeover from acceleration to deceleration. The people will become weightless for a few seconds and the seats will be turned around automatically. The people must be strapped into their seats during that time. The entire inside of the ship will become the mirror image as far as the passengers are concerned.

Prior to landing, the ship must turn around. Another period of floating will occur. All these periods should take only a few minutes. The crew of course will be quite used to weightlessness for short periods of time. However some will have to be trained to work under weightless conditions for emergency repairs and the like. The passengers will merely have to remain strapped in their seats during this time. A special room could be set up for those passengers who wish to experience weightlessness for a while.

The trips to the moon will become common place. It will take no longer to get to the moon then it does to go from New York to California. Therefore Hotels will be set up on the moon. Everything must be brought in but the proton thruster engines will enable solid rock to be cut and fused together. We will then be able to build domes and bring water and supplies with us. There will be constant daily trips to the moon. People will honeymoon on the moon as well.

We will develop space stations on the moon with sufficient supplies to keep several hundred people alive for up to one hundred years. Cargo ships will bring a lot of water to the moon and we will produce a small city in the rocks of the moon with sunlight coming in. We can then grow fresh tomatoes and other vegetables.

Section 10-4: Trips to Mars.

Mars is approximately 142 million miles from the sun. We are 93 million miles from the sun. Although orbits vary, we can assume that at some time we will be approximately 55 million miles Mars. Astronomers and the like know the actual distances. However for the discussion here, it is only necessary to understand the approximate distances. The time to reach 27.5 million miles at 1G would be 26.5 hours with a peak speed of 2.08 million miles per hour. The entire trip to Mars would take 53 hours or 2 days and 5 hours.

Of course, the exact time of the trip depends upon the orbit and the time of year. The trip to the moon could be a day trip. Many people can go. Millions of people can eventually make the trip to the moon each year. It will not be an expensive trip. The trip to Mars will be much more expensive for the average person.

Section 10-5: Trip to reach Maximum Speed

Our brave astronauts will have to test the ability of man to achieve high speeds and the ability of man to reach deep space. We must test the limits. We must find out how far we can travel from the Solar System and still survive. We will head toward outer space and turn back when necessary.

At the conservative 1G it will take us two and a half months to achieve 20% of light speed and another two and a half months to return to the Earth. The biggest problem besides fuel is the Doppler length distortions.

Let us look at the following chart for the variation of gravitational shrinkage and Doppler expansion with respect to increasing speed.

Velocity	L Grav.	L Rear	L Forward
0.0	1.000	1.000	1.000
0.1C	0.995	1.1	0.9
0.2C	0.980	1.2	0.8
0.5C	0.866	1.5	0.5

The Einsteinian gravitational length is the shrinkage of distance with velocity as per the space-time equations produced by Einstein. Unfortunately his relativity did not account for the Doppler differences. At the atomic level the gravitational length of particles contract and the length of particles increases in the rearward direction of motion and decreases in the forward direction of motion.

These changes also affect the binding energies of the atoms and compounds. This will cause corresponding changes in the large objects such as a spaceship. However the exact changes at the large scale dimensions will require future tests of high speed spaceships. These Doppler differences produce distortions which will tend to destroy living creatures.

These differences are not important at low velocities such as we have upon the face of this Earth. When we attempt to bring a spaceship upward toward the speed of light, objects get distorted. At fifty percent of the speed of light a 6-foot person becomes over 9 feet in the rearward direction but only 3 feet in the forward position. A ceiling of a spaceship compartment, which is twenty feet, becomes only 10 feet in the forward dimension. We must make sure the ceilings are at least 20 feet high between compartments on the spaceship.

When we look at the workings of our own body we find that our organs will struggle really hard when we attempt to reach 50% of light speed.

As we look at the chart we find that the particles within a spacecraft traveling at 0.1C have a slight Einsteinian shrinkage of one half of one percent. This is small. For the case where the spacecraft dimensions match the particle Doppler changes, the spacecraft looks ten percent smaller in front of it and ten percent larger to the rear of itself. A man would experience the same distortions.

When we go up to 20 percent of light speed, the Einsteinian gravitational shrinkage is 2 percent. The forward length is 20 percent less and the rearward length is 20 percent more. These distortions are getting large and the life of the crew is in peril.

As we look at the chart we see that it is impossible for us to achieve the star Alpha Centauri in our lifetime as long as our technology limits us to 0.2C. We are limited to our solar system. Unless we are able to travel into the Cs/Co dimension, this solar system is our final place of existence.

There are philosophical possibilities and scientific possibilities which could enable man to go beyond our solar system. In my philosophical books, I discuss various possibilities. In Chapter 16, I discuss one scientific possibility.

Chapter 11: The Light Wave and the Photon

Section 11-1: General Concepts

Let us now attempt to get a picture of the photon from a dual light speed Co/Cs perspective. At this point in the study, we know that the photon will absorb dot-waves when the gravitational field intensity increases and it will lose dot-waves when the gravitational field intensity decreases. As a photon approaches a star or planet, the intensity of the gravitational field increases and the photon turns toward the blue. When the photon leaves a star or planet the intensity of the gravitational field decreases and the photon turns toward the red.

When a planet moves toward a star, the intensity of the gravitational field increases and the photons it receives will turn toward the blue. When a planet moves away from a star, the intensity of the gravitational field decreases, and the photons it receives will turn toward the red.

The intensity of the gravitational field at any point between a planet and an object depends upon the mass of the objects, the mass of the planet, the square of the inverse distances, and the relative motion of the objects.

The question is what does the photon look like? We know that photons spin. We also know that photons move at a constant velocity Co in a vacuum. In addition we know that photons are composed of dot-waves and dot-waves move from the Co dimension to the Cs dimension and back to the Co dimension. We also know that they are electrically neutral and have no mass.

To have mass necessitates having standing wave patterns. The photon has no mass and therefor it has no standing wave patterns. The dot-waves are always moving with linear motion and planer motion in the perpendicular direction.

Since the photon has a neutral electric charge, it has to be made up of particles similar to an electron and a positron. The energy levels of most photons only are a tiny fraction of the electron/ positron energy combination. Some could also be much larger such as when electro-photonic energy is added at CERN to produce protons and antiprotons.

The only way that photons can oscillate between dimensions and move at light speed Co in a linear fashion is if the negative dot-waves are moving in the Co dimension while the positive dot-waves are oscillating in the Cs dimension. Then when the equivalent positron is moving in the Co dimension, the equivalent electron is oscillating in the Cs dimension.

The positron and the electron are always in the opposite dimensions. This also means that the electric and magnetic fields always exist in opposite dimensions.

Positive and negative dot-waves cannot destroy each other because they can never directly add to each other. If they existed in the same dimension at the same time, the net result could be zero.

We say that something is electrically neutral if it has a balanced blend of positive and negative charges. Yet the positive and negative cannot destroy each other. The universe cannot self-destruct to zero because the dot-waves can never achieve a state of zero.

The photon tends to travel in a straight line because it is stabilized in the Cs dimension. For the Co dimension, the radius of the photon in the perpendicular direction is rather small. However once the segment of the photon switches to the Cs dimension, the radius is large. Since half the photon is always in the Cs dimension, you always have a large radius oscillation and a small radius oscillation at the same time. This insures that the photon will tend to move in a straight line. In addition, the Doppler length to the rear of the photon is two wavelengths in both the Co and the Cs dimension.

When the photon passes a star, we get a strong gravitational gradient. This causes the balanced oscillation in the perpendicular direction to become distorted. The side of the oscillation facing the star will tend to shrink and the opposite side will tend to expand and this will turn the photon toward the star.

Einstein did very well in his space-time calculations for the location of particular stars during solar eclipses. In general his equations are good mathematical describing functions. However there is no such thing as a time dimension even though his equations work well. Thus his math is a good approximation to space-time but his physics is incorrect.

Section 11-2: The Red Photon

The Red Photon is the Red Hydrogen Faunhofer line of 0.6563 microns which has energy of:

(11-1) $E_{R.P.} = 3.0267E\text{-}19 = 1.8891EV$

This energy comes from the difference between the second Bohr Orbit and the Third Bohr orbit. The differential energy level is:

(11-2) $E_{B3\text{-}B2} = 1.89058 \ EV$

The Red Photon comes from the sun and the sun's gravitational field will reduce the energy of the red photon as it moves from the sun to the Earth.

The equivalent rest mass of the Red Photon is:

(11-3) $M_{R.P.} = 3.36767E\text{-}36 \ Kg$

Since the mass of a dot-wave is;

(11-4) M_D = 1.96867E-71Kg

If the energy of the red photon was converted into spherical standing wave energy, the equivalent number of dot-waves within the Red Photon is:

(11-5) #DW$_{R.P.}$ = 1.71063E35 dot-waves

The individual red-photon is linear and angular energy. It is not readily compatible to a spherical gravitational pattern. However when it is absorbed by an electron, it produces spherical energy patterns similar to the particles themselves. This occurs as long as the energy is confined to a common stationary point. If the point is moving some of the energy will become mass and some of the energy will remain as momentum. If the point approaches light speed Co, then most of the energy will remain as momentum.

There are an equivalent huge number of plus and minus dot-waves within the red photon. Since the photons are soft forms of energy and absorb and radiate energy no two red photons will have exactly the same number of dot-waves

Electrons tend to be harder forms of energy which form spherical standing wave structures. No two electrons will have exactly the same number of dot-waves but each electron will match other electrons fairly well. You cannot add two electrons together to get a stronger electron of twice the charge. However when you move an electron a balanced blend of plus and minus dot-waves attach to it.

Section 11-4 The Color of light with gravity

Although the universe from the absolute center point is expanding spherically at approximately the speed of light Co, the gravitational field of galaxies tends to be independent platforms. Multiple big bangs occurred on the spherical surface. This caused the gravitational field to appear as a sphere for the galaxy as a whole. We do not have a universe where the measured light relates to the absolute center of the universe. All we have is a universe where the speed of light is Co and the absolute speed of the photon measured from the absolute center of the universe is much higher. Thus:

(11-6) $C_{XYZ} = (Cx^2 + Cy^2 + Cz^2)$

When the speed on the surface of the universe is the speed of light in all directions, the total speed is:

(11-7) $C_{XYZ} = 1.732Co$

The physics of the universe is such that the expansion of the universe does not count and the rotation of the universe is one plane does not count. Finally the rotation of the plane itself does not count. Thus the only thing that counts is the velocity of an object with respect to a reference point.

The Sun is younger than the galaxy. The creation of the sun set the gravitation field relative to the sun. In addition the formation of the Earth set the Earth's gravitational field much younger than the sun.

In general the Earth itself is our Einsteinian independent platform. Thus upon the Earth the gravitational field is centered upon the Earth and it tends to be somewhat spherical although as the Earth spins around the sun we get complications. We then get a field that has orbital components.

When we look out to the far stars we are looking at the image of the past. The light reaching us has traveled billions of years to get here.

The photons leave the sun toward the blue due to the gravitational field of the sun. As it leaves the concentrated field it continually regenerates itself as it flows from the Co dimension to the Cs dimension. Slowly the light turns redder. Then it reaches the balance of gravities position between the sun and the earth and it looks like red/white. The photons then move toward the Earth and become redder. Finally the photons hit our atmosphere and start to look slightly white. Then there will be an increase of the Earth's gravitational field intensity and the light will look slightly bluer which makes it look pure white. Thus the color of the light and its energy level varies constantly with the gravitational field's intensity.

Chapter 12: Black Holes and other Interesting things

Section 12-1: Introduction

The Dot-waves produce sub-particles, protons, electrons, neutrons, and photons which appear in the linear portion of space-time. The universe has linear areas and quite a lot of non-linear areas. The non-linear areas of space-time can confuse people when they try to understand the universe in terms of linear theory.

Section 12-2 Black Holes

Black holes are one example. They look like areas of space right after the initial big bang which produced billions of mini black holes. These then exploded simultaneously. The galaxies were formed during the explosion of billions of these black holes.

The bits and pieces of the initial big bang were like a fireworks display in which the main explosion occurs and then a short time later the little bits and pieces of the explosion explode themselves. This produced a spherical surface with galaxies approximately equally spaced all around.

These multiple explosions do not have the same power as the initial big bang. The subsequent explosions occur at the time when the compressive photonic spherical momentum has already been moving outward. This helps to prevent the possibility of the complete destruction of the black holes. In fact if the black holes were completely destroyed the galaxies would not have a central point of reference and the universe would be a collection of stars in homogeneous patterns. We would not have galaxies if it were not for the black holes.

Stars and planets would exist but the universe would be a much different and less interesting place for the astronomers and people who like to look through telescopes. All you would have would be uniform distributions of stars. With no black holes there would be more stars and higher levels of background radiation. In fact without the black holes the levels of radiation may be so high that life as we know it would be impossible.

In effect the black holes are a means by which the surplus energy of the universe is stored and kept away from the outer stars and outer planets. Therefore life is not readily possible in a universe which is perfectly uniform and linear. If the oscillating universe could produce versions of the universe where black holes do not exist, then life would be unlikely unless some radioactive beings came into existence.

Although people think that black holes suck up planets and galaxies and photons and try to take over the universe, that is not exactly true. Black holes are not simple vacuum cleaners that just suck up the universe. Black holes are energy converters. Therefore they convert matter and photons into both light speed Co/Cs and Cs/Co dot-wave radiation.

In effect a black hole is a kind of food processor. It chops up the physical Co/Cs world that we see and measure and turns it into the world of light speed Cs/Co dot-waves that we do not readily see and measure. This enables the universe to perpetually exist.

At the time of the big bang, light speed Cs/Co energy was flowing toward a pinpoint. In the process, light speed Cs/Co energy was converted into light speed Co/Cs dot-waves. However the light speed Co/Cs dot waves continually oscillate between light speed Co high mass/constant energy and light speed Cs very low mass/constant energy.

As the forming gravitational pressure of a black hole exceeds a certain level, the ability of the Co/Cs dot-waves to oscillate between high and low mass becomes difficult. During the original big bang inversion, huge amounts of light speed Cs/Co energy were radiated. The big bang exploded completely because there was nothing external to the forming pinpoint. Thus at that point the gravitational constant approached zero. The result was a spherical surface of rapidly expanding light speed Cs/Co energy and light speed Co/Cs energy.

The black holes after the big bang had an external universe surrounding it. The gravitational constant surrounding the black hole increased. This slow increase of the gravitational constant stabilized the stars and the galaxies. Far into the future all the remaining black holes will explode as well as all the stars. Then the visible universe will be gone. Eventually all the particles will break apart and we will be left with a photonic Cs/Co universe.

As the universe expands, increasing numbers of light speed Cs/Co dot-waves will come into existence. This will increase the gravitational constant as the light speed Cs/Co pressure increases. This will cause the gravitational constant to increase toward infinity at maximum radius.

In some respects the combination of Co/Cs and Cs/Co dot-waves in space is similar to an atmosphere with air pressures depending upon the partial pressure of each gas in the air space.

As we look at the present black holes we see that the gravitational pressure within the center of the black holes is very strong. This is similar to the pressure in the ocean that keeps increasing as we go down miles from the surface.

The center of the present black holes cannot maintain the oscillation of the dot-waves from light speed Co to light speed Cs. The net result is that the pressure converts the mass of the dot-waves into basically massless photonic energy. The light speed Co/Cs and Cs/Co energy radiates from the black hole. Thus a point is reached where a black hole eats up the surrounding stars and planets and reaches a maximum energy density. At that point, the black hole radiates an amount of light speed Co/Cs and Cs/Co energy equal to the light speed Co/Cs energy that it eats up.

The light speed Cs/Co energy fills the universe with dark energy which expands the universe. This dark energy is now original light speed Cs/Co energy and when concentrated in space-time can reform into new light speed Co/Cs dot-waves. This can cause new black holes to reform in certain areas of space-time that appear empty to the telescopes. Yet they have enough energy to produce new galaxies in a short period of time.

There are many possible solutions for the universe. The universe could eventually convert all the dot-waves into light speed Cs/Co energy. This would head outward toward infinity but eventually it would compress and a new big bang would occur.

The universe could also have many dimensions. It could be a dual light speed multi-dimensional universe. Our light speed Cs energy could flow into other dimensions at slightly different Co light speeds and while we expand toward infinity they would have their own big bangs.

There are many possible answers for sure. The main effort of this book is to understand the dot-waves. Prior books have discussed a universe of only light speed Co dot waves and coexisting universes of higher and higher light speeds. This present study looks at the possibility of a dual light speed oscillating dot-wave.

The black hole is the place in space time where you have concentrated center points of a huge amount of Co/Cs dot-waves. This is the place where matter and photons are turned into spherical light speed Cs/Co dot-waves. The other side of the black hole is the light speed Cs/Co universe. Thus light speed Co/Cs energy enters the black hole, flows through a worm hole, and radiates Cs/Co energy into the light speed Cs dimension.

Scientists ponder the destruction of intelligence entering the black hole. However any intelligence entering the black hole will not be lost. It will exit the black hole and enter the light speed Cs/Co universe. Nothing is really lost in the universe. The intelligence we generate enters the higher light speed universe. At the next big bang this intelligence will flow back to the next universe. Therefore the structure of life and man will continually be reproduced forever.

To the astronomer it would appear that the black hole is a huge entity. Today scientists are measuring the gravitational field of black holes when two black holes revolve around each other. There are a lot of scientific jobs and theories to be done in the future to accomplish a final understanding of the physics of the universe.

Some scientists are trying to build mini black holes out of laser beams. This is a possibility if all the linear photonic energy is made to form a spherical energy field and an oscillating spherical magnetic field is added to it. This would involve three dimensional patterns of the photonic energy set upon a focal point. The result would be the conversion of linear energy into patterns of spherical energy.

With enough energy flowing into the focal point, the mini black hole should have developed and the output would be photonic waves of both light speed Co/Cs and Cs/Co caused by the destruction of the protons. This will provide us with abundant energy for the future. The downside is the creation of more powerful weapons of mass destruction.

Section 12-2 Continuous Creation

The universe is a complex entity. There are many solutions and within each solution there are areas of space/time in which non-linearity will cause differences to occur. Continuous creation can occur for the situation where energy flows from one dimension of the universe to other dimensions.

A dimension exists where the dot-wave oscillation has different top and or bottom light speed levels. If the lower light speed level Co is part of a spectrum of light speeds, many different dimensions can exist. If the upper light speed level is a constant for all dimensions, the intelligence from our dimension can flow into another dimension.

The expansion of our universe toward infinity can result in the creation of another universe heading toward a big bang.

Even in our universe a black hole can form from energy from another dimension. This can explode and form a new galaxy out of what looks like pure empty space. Since the speed Cs is so large, an area of pure empty space can produce a galaxy in a short period of time.

Even in our dimension, the destruction of a black hole can produce so much light speed Cs/Co energy that an empty area of space can fill with this energy and a new galaxy can form.

Section 12-3 Fast rotating stars

Some binary stars are spinning at a very fast rate. What holds them together? One solution is that the stars are held together by baby black holes. This will produce tremendous gravitational forces. Once we add Doppler photonic forces we end up with a situation which brings us close to the original big bang. The light speed Cs/Co radiation will be very high and thus the force of gravity will be much more than in linear space time. In addition as the binary stars expand apart from each other, other Doppler forces come to bear. Over time the stars will be eaten up by the baby black holes and the baby black holes will radiate away into the light speed Cs/Co universe.

We exist in linear space time. Our protons are expanding. The distance from our Earth and our Sun is expanding common mode due to the expansion of space. The forces are modest. Once we move into non-linear space time, modest forces can become tremendous. Therefore there are many possible reasons for the rotation of the stars.

Section 12-4 Electrical Radiating binary stars

If a black hole exploded not too long ago, we would get new protons and new electrons at a level of charge and energy not too far from the energy and charge right after the big bang. This means that the stars will be composed of high energy high charge protons and electrons.

We could even have a situation where nearly original super protons and super electrons still exist. They would have the same mass or nearly the same mass as each other. They would also have huge charges similar to the charge at the time right after the big bang. This can only occur in a very non-linear area of space time. This would cause a period of intense radiation of very high energy photons.

Some star and star systems could even spit out very heavy radioactive particles which quickly break apart when they reach gravitational fields in linear space time.

134

We would then get strange particles which used to exist long ago all over the universe but which only exist in non-linear space time. We find these particles as they are sent to us in explosions.

For the most part the high energy stars most likely have super protons and super electrons which have mass to charge ratios similar to our particles. However the dot-waves can produce a large assortment of interesting things in non-linear space time. The high mass stars will radiate at a higher energy level than ordinary photons.

We can also have super electron and super proton mini black holes within the electrically radiating stars. Excessive layers of positive dot-waves can be mixed with photonic balanced layers within the black hole which would give the black hole a huge equivalent charge. The sister black hole within the sister star would then be drawn to the other star with huge electrical forces. Therefore you can get tremendous electrical fields radiating outward in the universe.

There are a huge variety of particles and stars that can form in areas of non-linear space time. The gravitational and electromagnetic fields will work basically the same but the measurements will all be different than we observe here.

Section 12-5 Neutron Stars

Neutron stars are part of the physics that we see and measure. Astronomers and physicists have produced equations that represent the collapsing of regular stars in linear space time. The result is neutron stars. From a Doppler space-time perspective ordinary stars have particles with velocities which have motions which produce differences in frontal and rearward masses and associated Doppler photonic forces. This builds up photonic pressures which insure that the stars will remain stars. This is usually represented by heat theory which is the easiest way to express it. Electrical theory is fine but Mechanical theory is often much easier to use in many circumstances. In fact any attempt to write electrical type equations for the stresses on steel beams in construction would be meaningless. Since the mechanical equations and the electrical/mechanical type equations are both valid, whatever is easier to use is the best.

Scientists have produced formulas which show certain points where a star can form into a dwarf star. The radiation of an exploding star reduces the Doppler photonic energy to a very low level. Without the motion of the atoms, the photonic forces are eliminated. Then there is nothing to prevent the electrons from becoming part of the protons. In simple mechanical terms once the heat is radiated away a point is reached where the star will collapse. The remaining energy forces the electron into the proton and a neutron results.

Not only are the electrons and protons turned into neutrons but the neutrons have no kinetic energy left. The dot-wave radiated gravitational forces then brings the entire star into a small volume. The dwarf star may still have a little energy left and radiate. The dot-wave radiation under the influence of the strong gravitational field may still produce some ordinary photons. Although the neutron star is in linear space time the star itself is quite non-linear and the surrounding gravitational field is also non-linear.

The neutron star still maintains the neutrons. Over time as more and more spherical dot-waves within the neutrons flow into the stars center, a mini black hole can form.

Section 12-6 Non-linear space time

The model of the universe is based upon linear space time. This is a best fit approximation to the universe. In reality there are sections of the universe where linear space time is applicable. In addition there are other sections of the universe where non-linear space time is the rule. This means that it is impossible to accurately represent the entire universe in terms of linear space time. The simple model of the universe will appear quite different in many places within the universe. The dot-wave model is a simple three dimensional picture of the universe. When the light speed Co and Cs dimensions are added, the universe will still look the same in the linear regions. Then all we have are various non-linear regions to observe and attempt to understand.

Non-linear regions can produce sections of the universe where the gravitational field twists and turns into strange shapes. Some galaxies will exhibit strange things. This will cause the astronomers to puzzle at the interesting things that they see.

As an Engineer I am happy with simple models of the universe. This enables me to fairly understand how the universe works. My simple model of the photon gives me a visual object which I can study. Mathematicians and scientists look into the details of the way particular things work. They study the stars in great detail. I am only looking at the way the stars work. The exact details are not my concern or ability. I am looking at the forest while others are studying each tree.

Scientists have great labs and huge amounts of computers and associates. All I have is pen and paper and a $15 calculator. It is not possible for me to study the exact details of the stars and all the things that exist in the universe. I cannot work with multi-dimensional math. At 79 years I can no longer do complex integrations that I did when I went to school long ago. I can no longer do a Fourier series. All I can do is simple algebra and very simple calculus. However the dot-wave theory itself is rather simple from a linear perspective. Non-linear dot-wave theory is beyond my mathematical ability. However all I propose in this book is an Engineering solution to the universe. The complexities of non-linear dot-wave theory are left for the future.

Section 12-7 Jumping Particles

Quantum theory uses probability to discuss interesting properties of the universe such as the double slit experiment. They have performed experiments which showed the ability of particular particles to interact with sister particles hundreds of miles away much faster than the speed of light.

The electron has the property of being a particle one second and a wave another second. The double slit experiment shows this effect. You cannot really guarantee that the electron will remain a particle at any particular instant in time. If you have a beam of electrons hitting an old fashioned TV tube, there is a probability that most of the electrons will behave like particles. Most electrons will, otherwise the TV picture would be poor. However many electrons will not remain particles and will merely become part of an electron photonic charged wave.

The electron is composed of only negative dot-waves. Therefore the electron is a synchronized device. All the dot-waves exist in the light speed Co dimension at the same time and switch to the Cs dimension at the same time.

When the electron moves into the Cs dimension there is no guarantee that it will obey the laws of particle physics. In the double slit experiment, the electron did not obey the simple laws of particle physics. This is because it was free to move huge distances at will.

When the electron is in the Co dimension is a combination of a stationary/ moving wave. When it is in the Cs dimension the same is true but it can move at light speed Cs. Thus there is the probability that sometimes the electron will stay where it is and sometimes it will travel huge distances.

In general all moving electrons have pure negative dot-waves and a mixture of photonic plus/minus dot-waves. This helps to prevent all electrons from moving huge distances at will. In the Co dimension the electron sees the slits for the double slit experiment. However in the Cs dimension things are different. The distances involved are much larger. There is a probability that the electron will not see the slits. Thus the wave passes through both slits because they do not exist all the time for the Cs dimension.

The electron can jump from one spot to another. Heisenberg's uncertainty principle specifies that the simple laws of classical physics cannot be used to perfectly define both the position and velocity of a moving electron or object at the same time.

One problem is that time is not a dimension. It is just a measurement. Another problem is an electron exists in the light speed Co dimension one split microsecond and then exists in the Cs dimension another split microsecond.

Therefore there is no simple relationship as the electron exists in one dimension and then moves to the other dimension. The universe is different when viewed from either dimension.

For years I tried to explain the universe in terms of a single light speed. I never was satisfied with the results.

The dual light speed universe makes explanations much easier to understand. Instead of trying to understand the uncertainty principle to little avail, it is clear that the transition between dimensions produces an uncertainty or jitter.

We say that the electron is both a particle and a wave. What does that mean? A particle is a situation where you have standing wave patterns. A wave is a situation like light where you have moving wave patterns. A moving electron is both. It is a combination of standing waves and moving waves.

In some respects this is similar to a Fourier series. There is a fundamental frequency and a series of harmonics. The standing wave could be represented by the fundamental frequency and the moving wave could be represented by the series of harmonics.

I do not know if similar things are used in wave mechanics since I avoid complex math. To do complex math you need a good memory. I have a good calculating brain but my memory abilities are quite limited. In addition my engineering mind is more interested in how things work and building things. I am content to understand simple algebra and simple calculus. When I was younger I was much better at understanding more complex math but that was a long time ago.

In order for me to understand the waves, I merely put a picture of a standing wave and a moving wave in my mind. Then it becomes clear to me that the property of mass is standing waves and the property of momentum is moving waves. The complex wave equations are not important to me for an engineering level analysis of wave mechanics.

The gravitational and electromagnetic fields are composed of free dot-waves within space itself. The dot-waves of the field vibrate from the Co to the Cs dimension as well. An intense field will have a high density of dot-waves per cubic meter and a weak field will have a low density of dot-waves per cubic meter.

The gravitational and electromagnetic fields have mass. A moving particle hits the mass particles within the field. Yet there is no exchange of momentum. Any change of momentum requires minimum quanta of mass/energy to have an effect. Thus a particle can hit the fields without any effect.

This is probably due to the fact that we are dealing with waves rather than balls of energy. We look at a proton and say that it is a ball of energy. That is a describing function analysis. The reality is much more complex when we look at the proton as a structure of complex waves.

When we look at an ordinary ball, it looks like a spherical object to us. Yet if we could look inside the ball it would be a very complex structure of standing waves and moving waves.

This is similar to the structure of our own bodies. If we could look inside us we would find strange parts. Then when we look inside the parts they would appear even stranger. If we tried to write equations for the human body, it would take billions of billions of equations that no one could understand.

The net result is that we have describing functions which explain how a ball moves when hit with a bat. We do not write complex wave equations to describe this. For the most part things are described from an engineering type viewpoint and not a complex scientific viewpoint.

Quantum mechanics specifies that an electron can be here one second and a hundred miles away in another second. This is not quite easy for a proton to do since the proton has a much more complex structure. The electron being all negative dot-wave electrical energy can readily synchronize and become part of the high light speed Cs universe and then reappear elsewhere.

The same is true of various sub-particles which have relatively simple structures. The ordinary proton has a very complex structure and it would be difficult for a proton to travel huge distances like the electron can. Thus the probability of a proton moving large distances would be very low but not impossible.

An electron at one location can travel at light speed Cs and produces another electron far away. It is thought by people who take Quantum theory to the limit that a machine could be made to transmit a person a hundred miles. The problem is that the probability of moving a single synchronized electron is a quite high but the probability of moving huge numbers of electrons and protons is basically zero.

What is the probability of a structure of a limited amount of dot-waves jumping a hundred miles? The answer is that it is easier for a low energy dot-wave structure to jump a billion miles then it is for an electron to jump one hundred miles. In fact there is no certainty that you can define the position and velocity of a low energy structure of dot-waves.

The wavelength of a single dot-wave is compatible with the radius of the universe. A single dot-wave has no certainty of existing anywhere. We have to add billions of billions of dot-waves together to produce an electron. Then we have a degree of certainty that the electron will exist. The proton has a greater degree of certainty and a spaceship has a very high degree of certainty.

The basic structure of the universe is such that zero is the inversion of infinity. The size of a dot-wave is basically zero while the universe approaches infinity. From this is appears that everything in the universe has an inverted image.

Everything in the universe jumps from position to position because they are moving from one dimension to another dimension. As they jump between dimensions and return to their starting dimension there is an uncertainty of position and velocity as Heisenberg discovered.

Section 12-8 Space

What is space? It can be described as something that carries the waves. It is something that converts electron flow into magnetic fields. It is something that expands or contracts. Thus it is a very flexible entity. Some people feel that it is a hard substance harder than a diamond.

Now let us look inside of space. We find gravitational waves and fields. We find photons. We find electromagnetic fields. We find single vibrating dot-waves. We find light speed Co/Cs photons and light speed Cs/Co photons. We find Co/Cs negative parts of photons and Co/Cs positive parts of photons. We find Cs/Co photons as well. Thus there are huge amounts of high light speed mass/ energy in space. For possible coexisting universes we find a whole spectrum of other combinations of light speed for other dot-waves. Thus space it is a very busy place.

The question is whether space exists by itself? We have been taught that the impedance of space is a measured value. We have come to accept that space contracts around a star and this causes the light to bend. Does any of this make sense?

Let us now remove the visible material, the photons, and everything else from space. In addition let us remove the light speed Co/Cs photons and light speed Cs/Co photons and all electro-photons from space. What do we have left? A simple equation explains it all.

Equation (12-1): Space = 0

Equation 12-1 tells it all. Space is nothing at all. Space itself is made from nothing at all. Space itself has no structure. Space has no properties at all. Space cannot be elongated or compressed. Space cannot be filled with tiny dimensions such as string theory presents.

The greatest scientific minds have attempted to define the properties of space. Yet it is self-evident that there is no such thing as properties of space.

When the entire universe shrunk to a very small size, there was nothing outside of that size. Everything that existed was inside the small size. It may be possible but not necessary for other universes to exist in the same space we occupy. Aside from that possibility, there is nothing beyond the distances of our universe. What we see is all there is. It is what it is and there is nothing else.

There is no answer for how the universe came into existence and there is no answer for what lies beyond the universe. Everything is self-contained within the universe. Space like time is meaningless. Time is just a measurement and space is just a measurement

What we consider the properties of space are really the properties of the dot-waves themselves. We live inside an expanding ball of energy of light speed Co and light speed Cs. The properties of space are the properties of this expanding ball of energy.

The physical universe is a huge energy sphere of radius 2Ru in the Co dimension as illustrated on the picture of the Universe. Outside this sphere is a much larger sphere of light speed Cs energy and everything that we call space is really inside the sphere.

The gravitational dot-waves in dark matter/dark energy provide us with the basis of the gravitational and electromagnetic fields. When Einstein says that space shrinks he is mistaken. It is the fields that shrink.

When we say that space has impedance, this is incorrect. It is the fields that have impedance. The distance between the Earth and the sun is measured along the fields. The fields have definite spacing and everything jumps from one field line to another as they oscillate from the Co dimension to the Cs dimension and back.

Chapter 13: The Multi-light-speed universe

Section 13-1 Basic Concepts

The visible universe we live in exists on an approximate spherical surface at a distance of Ru from the absolute center of the universe. At the same time every point in the visible universe is at the center of the invisible universe which extends from the absolute radius of Ru equals zero to a radius 2Ru. Ru is a variable radius which expands from near zero at the big bang inversion toward near infinity at full expansion.

The visible universe is expanding and over time it will erase and be converted into light speed Cs/Co energy. Far into the future another big bang will occur and our light speed Co/Cs universe will return.

It is possible that our universe is part of a more complex structure. It is possible that we live in a multi-light-speed physical universe. The interconnection between these universes is the high light speed universe of light speed Cs. The Dot-wave theory goes beyond just our universe and looks at the entire spectrum of possible higher light speed physical and perhaps photonic universes.

To us the original big bang appears as a singular event. After the singular event the universe stretched rapidly out at a light speed near the geometric mean of Co and Cs and upon a spherical type surface where billions of other little bangs occurred. The big bang was the inversion of spherical Cs/Co energy into Co/Cs energy.

A particular solution for the big bang would be our universe alone. The general solution is different. For the general solution there would be a spectrum of inversions. This would cause a spectrum of universes to form. There are many possible solutions to this series of universes.

146

One solution would be a series of universes in which the light speed Co of each visible universe is double the light speed of the previous universe. If all inversions occurred near the same time, we would get a series of light speeds such as 0.25Co, 0.5Co, Co, 2Co, 4Co, 8Co, etc. If the distribution of dot-wave energy was constant, we could produce a chart of the dot-wave characteristics for each universe.

Light-speed	Charge	Mass	Energy	Radius
0.25Co	$4Q_D$	$16M_D$	M_DC^2	0.25Ru
0.5Co	$2Q_D$	$4M_D$	M_DC^2	0.5Ru
Co	Q_D	M_D	M_DC^2	Ru
2Co	$0.5Q_D$	$0.25M_D$	M_DC^2	2Ru
4Co	$0.25Q_D$	$0.0625M_D$	M_DC^2	4Ru
8Co	$0.125Q_D$	$0.0156M_D$	M_DC^2	8Ru

The above chart is for the spectrum of universes in which the dot-wave energy is constant and the product of the dot-wave charge and the light speed is also constant. The lower light speed universes are closer to the absolute center of the universe. The upper light speed universes are further away from the center than us. All the dot-wave fields intercept the absolute center of the universe and each universe also intersects a sphere at twice the radius of their physical universe.

The visible universes do not coexist. They are different spherical planes separated by large distances. All the dot-wave fields do coexist with each other

We can also go beyond the light speed Cs dimension and move toward light speed infinity. As the light speeds go up, the dot-wave masses decrease as the inverse square of the light speed. When we move upward high enough, the mass is so low that no protons and electrons can form. All we get is photonic fields. At light speeds near infinity, the photonic fields are extremely stretched out.

Multi-light-speed physics involves many philosophical possibilities. The universe could be merely a machine which stores intelligence and feeds it back into the forming universes producing man and life forever. Alternately the higher light speed universes could be looked at as producing higher versions of man and life.

In addition there could be upward mobility as the prior universe dies out and man and life move upward slowly toward light speed infinity.

Chapter 14: The Gravitational Field

Section 14-1 Introduction to the Gravitational Field

There was a time prior to the big bang, when the entire light speed Cs/Co universe compressed and flowed through a common worm hole. The compressing universe appeared as spherical planes which oscillated through the worm hole with every decreasing size and ever increasing frequency. As the light speed Cs/Co dimension shrunk toward a minimum radius a point was reached where the size of the radius was:

(14-1): R= Rmin

At the time where the Co/Cs universe was starting to form, the outer radius was:

(14-2): Rmax = 2 Rmin

We ended up with a shrunken universe that looked like the picture of the universe but much smaller. The explosion caused it to increase in size but the picture remained the same.

The universe expanded rapidly at a transition speed. The speed of expansion is the geometric mean Cgm of the light speeds Co and Cs. Thus:

(14-3) Transition speed = Cgm =$(Co\ Cs)^{0.5}$

Equation (14-3) tells us that as the high energy dot-wave spheres moved from the Cs/Co universe to the Co/Cs universe and the light speed went through a transient condition. During this time the geometric mean speed more accurately shows the expansion velocity. This is especially true because there are many spherical layers of energy inverting at the spherical surface at the minimum radius Rmin.

At the big bang inversion, beyond the volume at 2Rmin, another spherical boundary was forming. The dimension of this outer sphere is:

(14-4) Rs = (Rmin)Cs/Co

The outer sphere is the minimum radius right after the universe stabilized time the ratio of the speed of light Cs to the speed of light Co.

Beyond this for the simple dual light speed solution there was no space, there was nothing at all. Space is only a measurement of the gravitational field. Everything that existed was inside the final sphere or radius Rs.

The inversion caused the Co/Cs universe to expand rapidly to a surface far away from the initial inversion sphere. The surface of the forming Co/Cs universe then exploded into billions of black holes. As the universe expanded its volume grew and what we call space developed.

Along a spherical surface at the midpoint of the expanding bubble of energy, the visible universe formed. Billions of mini-black holes exploded simultaneously and the visible universe took shape.

Visible energy radiated outward into the forming universe. We ended up with a situation in which the visible energy of the light speed Co/Cs universe existed at the distance Ru from the absolute center. The electromagnetic and gravitational fields of the universe existed from the absolute center to the distance Ru and also existed from the distance Ru to the distance 2Ru. At the distance 2Ru, the outer Co sphere has spherical surface forms of Co/Cs energy which oscillate to Cs/Co energy. In the far future all the Co/Cs energy in the visible universe will be gone and the outer Cs shell will rapidly flow at light speed Cs back toward the pinpoint. This will mix with the invisible remnant of the Co/Cs universe and the mixture will flow toward the big bang at the geometric mean speed. Then the next big bang inversion will occur.

We then have a continuous oscillator. The universe will rapidly expand to a large radius at the transition. It will then follow an exponential e^x curve at it expands. At the same time the mass of the physical universe will decay as an e^{-x} function. Then the universe will shrink at light speed Cs and then Cgm toward the pinpoint.

Although we used a sinusoidal solution for the cycle time of the universe, the most likely solution is the exponential. After an initial rapid expansion at velocity Cgm after the big bang inversion, we get a variable time for each segment of the Cs/Co waveshape with a geometric mean of Cgm. This looks like an initial impulse function to us for an electrical analogy. Then the forward expansion time will be slow whereas the contraction time in the far future will be very fast at velocity Cgm. This will look like another impulse function to us.

An electrical analogy of the impulse spikes of the universe is a current spike of energy which charges up a capacitor. This is followed by a current source which slowly charges the capacitor. Finally we get a current spike which discharges the capacitor in the far future. This repeats forever.

The question posed by this chapter is what is the gravitational field composed of? To answer this question let us look at the Sun's gravitational field.

Section 14-2 The Sun's Gravitational Field

The sun was formed several billion years after the inversion. It was formed from surrounding energy. The sun radiates linear photonic energy in the form of light and a spectrum of other radiation. It produces light waves.

151

It also expands and radiates both light speed Co/Cs and Cs/Co energy. The protons and electrons lose energy and expand over time.

The loss of energy and also charge from the particles and sub-particles within the sun form both the gravitational and the electro-magnetic fields. The dot-waves can be individual but for the most part they exist in groups.

The gravitational field is composed of balanced blends of plus and minus dot-waves. All matter radiates dot-waves resulting in a loss of mass/energy per unit time.

(14-5) $F = Co\ dM/dt + Mo\ dC/dt$

The force of gravity acting upon an object is light speed Co times the derivative of mass with respect to time plus the initial mass Mo times the derivative of the speed of light C with respect to time.

The first term is the radiation of dot-waves which leave the Co dimension and flow into the Cs dimension. They are only a small part of the Co/Cs oscillation. The second term looks at the original mass Mo and the change of light speed from Co to Cs.

This equation attempts to look at the problem from ordinary physics. However what is happening is not ordinary. Mass and energy is radiating into the Cs dimension with a resulting back pressure which is gravity.

The gravitational field is similar to an electric current flow except that both positive and negative dot-wave currents cause balanced fields which attract each other.

Two transmission wires will attract each other when the current flow is in the same direction. When the electric current is in the opposite direction the wires will repel each other.

It would appear that the balanced plus and minus dot-wave currents of one mass would not attract another set of balanced currents from another mass.

We know that G and Uo have the same units. Therefore the gravitational field and the magnetic field tend to be equivalent. We also know that the positive dot-wave charges do not occupy the same space that the negative dot-wave charges occupy. This makes it appear that the oscillating positive and negative currents act like they are in different dimensions.

From a mechanical perspective the force of self-gravity is the loss of mass per unit time. Thus a ball of mass is held together because the waves are radiating outward from the mass. If we take two masses side by side, the gravitational fields of the two balls of masses tend to form a combined spherical surface at a distance.

The gravitational field is vibrating. After the Co/Cs dot-waves have become Cs/Co photonic waves, their mass has been reduced to a small fraction of what it was. The loss of mass of the ball is equal to the original Co/Cs mass less the mass of the Cs/Co dot-waves.

Normally for the Co/Cs universe the mass oscillates between the very high mass low light speed state and the very low mass high light speed state. There is no loss of mass for most of the dot-waves. The mass oscillates but it is not lost.

The very small percentages of the radiated dot-waves which permanently move from the Co/Cs universe to the Cs/Co universe make up the gravitational field. The mechanical perspective obeys the standard laws of classical physics.

Now we have to understand the forces from an electrical perspective. Something different is happening. For the Co and Cs dimension, the plus and minus dot-wave currents which flow between the dimensions tend to be equal and opposite.

Plus dot-wave currents flowing in the same direction as other plus dot-wave currents tend to attract each other. Minus dot-wave currents flowing in the same direction as minus dot-wave currents tend to attract each other. Plus dot-wave currents and minus dot-wave currents flowing in the same direction as minus dot-wave currents have the same effect as plus dot-wave currents flowing in the opposite direction as other plus dot-wave currents. The net result of all the current flows within the Co and Cs dimensions and through the worm holes between the dimensions have zero net attraction or repulsion.

The thing that is different with the gravitational field is that there is a permanent discharge of the plus and minus charge Q in the Co dimension due to the radiated dot-waves. In normal electrical theory we do not have a loss of the charges. For a simple L/C electric circuit, the charge of a capacitor can decrease while the current through an inductor increases. The energy has changed from stationary charges to moving charges. Then the current flow will decrease and the charge of the capacitor will increase.

The gravitational field is different. When we look at the dual dimensions of the universe as two different electrical circuits, the L/C oscillation in the Co dimension looks like an ordinary electrical circuit for most of the dot-waves. The same is true when we look at the oscillation in the Cs dimension.

The problem is that the energy of the Co dimension is going down while the energy in the Cs dimension is increasing. Thus there is a transfer of energy between the dimensions. The net result is that although plus dot-wave current and minus dot-wave current within a dimension

154

tend to balance out with no net attraction; this is not true for current flows between dimensions.

When we studied the expansion of the hydrogen atom, there was a force between the normal Bohr orbit current flow and the radiated current flow.

It then becomes clear that a force of gravity exists between the normal dot-wave positive current flow and its radiated component as it crosses the barrier between the dimensions. The same is true of the normal negative dot-wave current flow and its radiated component.

The normal flow is large but the radiated component is very small. This causes the force of gravity to be a magnetic electrical force between dimensions. The normal electrical forces are within the same dimension whereas the gravitational electrical forces are between dimensions. That is why the gravitational electrical forces are so weak.

In electrical theory we have a point charge which has an electric field surrounding it. Maxwell's equations discuss point electrical current sources. These are related to the magnetic fields. Yet this is all within the light speed Co dimension. In effect the magnetic field is creating the current flow.

These are large scale current normal current flows. The current flows between dimensions are true point current sources because they come through the worm holes. They are microscopic current flows as compared to the normal charge Q current flows.

We then have to produce Maxwell type equations for currents and voltage fields between dimensions. There is a lot of future mathematical and scientific work to be done in this regard.

In any event the gravitational fields are magnetic fields caused by current sources which are balanced one way current flows between dimensions.

The mechanical equations that are used are the loss of mass per unit time while the electrical equations that are used are similar to classical electrical equations except that equal amounts of plus and minus dot-wave current flows do not balance out to zero.

As we look at a photon, it tends to spin. Since it is electrically balanced as an equal number of plus and minus dot-waves, it would not appear magnetic. However the plus and minus dot-waves oscillate between the Co and Cs dimensions such that they never occupy the same dimension in the same state at the same time. Thus the plus dots at light speed Co exist when the minus dots are in the Cs dimension and vice versa.

They cannot destroy each other because they never coexist in the same place and the same time. The net result is that they appear electrically neutral from a total charge perspective. On the other hand from a magnetic field perspective they are not neutral. This gives us the property of spin.

The net result is that two objects have spinning magnetic fields due to the interactions between the Co and Cs dimension. These fields are quite complicated and the math has to be written in the future.

The net result is that the gravitational field between the sun and the Earth is magnetic. The Sun's gravitational field interacts with the dot-waves within space to produce a field which extends huge distances. There are interactions which occur at light speed Co. There are also interactions which occur at light speed Cs.

When we study the stars and planets we cannot view interactions between them as only caused by light speed Co. We most look for higher light speed interactions at light speed Cs.

Since G and Uo have the same units, the gravitational field is a different type of magnetic field. It is interesting that the photon has a measured spin but zero magnetic moment. The photon is electrically balanced in the amount of plus and minus dot-waves.

It seems to be balanced in terms of magnetic fields as well. However the photon is radiating light speed Cs/Co dot-waves. This produces a very high light speed magnetic field component. It may very well be that the net magnetic force of gravity takes place in the light speed Cs dimension. This would make it very difficult to measure using modern instruments. In the future it may be possible to devise indirect light speed Cs instruments that pick up the light speed Cs dimension magnetic moment.

Section 14-3: The Gravitational Wave

We cannot just look at the physics of our light speed Co dimension alone since the Cs dimension insures that the gravitational laws work. We can only readily see and measure part of the total universe. We say that space has certain properties. However that is not true. It is the interaction between the Co dimension and the Cs dimension that makes things work.

When we try to push an object it resists us. This is because as we push against the mass in the Co dimension, the mass in the Cs dimension is stationary. We then have to add photonic energy to push the object. This is needed to overcome the reaction from the Cs energy of the object.

157

In effect the property of inertia is an inter-dimensional interaction. This is because the dot-wave is not an ordinary electrical device as such. It is a dual light speed device. Energy from our universe at our light speed passes a pinpoint or spherical surface and flows into another universe at higher light speed. Einstein would call it space-time. That is how he defined it. In reality it is a dual light speed problem where instead of the time dimension we have the light speed dimension. Yet the equations are quite similar.

Since light speed is meters per second, if we increase the light speed while keeping our ruler constant, we can say that our time dimension changed. Then Einstein would be correct to say such things. For me it is clear to think in terms of light speed changes rather than changes in time.

The sun's gravitational field has traveled several billion light years since it came into existence. It does not exist beyond that point. There are many galaxies that are very old. Their gravitational field exists for perhaps 13.78 billion light years in distance. Thus our sun's gravitational wave is small in size compared to our galaxy as a whole and the other galaxies.

When we look at the sun's gravitational field in detail, we cannot find the field by itself. The dot-waves of the field combine with the dot-waves of all other stars. The net result is that at a great distance you cannot readily find the component of our sun's gravitational field.

Our Earth has a gravitational field composed of patterns of dot-waves. The same is true of the other nearby planets. If you took a measuring instrument in nearby space you would be able to detect the Earth's field for a reasonable distance. However as we move farther away all you get is a combined gravitational field from the sun and all the planets.

When the sun was first formed, the gravitational field had a transient condition in which a strong wave occurred. Thus when all the dot-waves are forming patterns in space astro-scientists could detect the individual gravitational field of a forming star.

The photons enable us to see stars far away. The structure of the photon enables it to produce light waves produced by billions of billions of photons. The stability of the individual photon enables it to survive billions of years.

The transient gravitational wave will be measureable for quite a long time when the intensity of the gravitational field is very strong. Thus scientists are able to detect gravitational waves from rotating black holes. This is similar to a gravitational wave from exploding stars. However visible light and X-rays are much easier to detect than gravitational waves from the formation or destruction of ordinary stars.

Chapter 15: The Electromagnetic Field

Section 15-1: The Electric Fields

The electromagnetic fields are formed in a similar manner as the gravitational fields. The radiation from particles produces positive dot-waves, negative dot-waves, and bipolar dot-waves. If we split the gravitational field into pieces we will get the electric field.

If we combine positive and negative particles such as an electron and a positron we will get photons. If we pump enough positive and negative energy into a small area we will produce protons and anti-protons.

Thus the mechanical world can produce the electrical world and the electrical world can produce the mechanical world. We can put mechanical power into a generator and turn out electrical power. Likewise we can put electrical power into a motor and produce mechanical power.

Energy is the same whether mechanical or electrical. That is why we can describe the universe as an electrical universe or a mechanical universe. The conversion is:

(15-1): Mass = Charge x Light Speed

Equation 15-1 is a repeat equation from the mass to charge velocity chapter. When we look at the fields we find:

(15-2) Gravitational Field = Magnetic Field

Equation 15-2 specifies that the gravitational field and the magnetic field are the same type of field. The gravitational field requires an interaction between the dual light speed dimensions. The ordinary magnetic field operates in each dimension. However there are components of the ordinary magnetic field in both dimensions. The same is true with the electric field.

The fields are the results of huge amounts of distributed dot-waves in space. Space has no property by itself. It is the dot-waves within space that produces the properties of space.

As the particles radiate plus and minus dot-waves in space they tend to interact with each other. The positive dot-wave oscillates from the light speed Co universe to the light speed Cs universe. At the same time the negative dot-wave oscillates from the light speed Cs universe to the light speed Co universe.

The positive and negative dot-waves occupy different dimensions. They attract each other but they cannot destroy each other.

The electric forces are the result of powerful space time forces between positive and negative dot-waves. Gravity is a force between balanced pairs of plus and minus dot-waves which interact between dimensions while the electrical force is a force between the plus and minus dot-waves themselves in the same direction.

Studies of photonic beams flowing in the same direction do not attract each other. However photonic beams flowing in the opposite directions do attract slightly.

The scientists have not been able to measure any magnetic moment from photons. Since the photons are very similar to the spherical gravitational photonic fields, any measurement of their magnetic moments would not be noticeable at light speed Co. The forces involved are gravitational and not standard electrical. Therefore these forces are tiny as compared to regular electrical forces.

The dot-waves that are radiated from the Co dimension to the Cs dimension are such that plus dot-waves and minus dot waves flow in the same direction. That is equivalent to positive dot-waves flowing in one direction and negative dot-waves flowing in the opposite direction.

Since the photons only respond when the beams are in opposite directions, it may be reasonable to assume that the gravitational dot-wave flows are not equal and opposite and thus the net result is an attractive force.

The normal electric forces only works because of the Co/Cs space time oscillation. A plus charge attracts a minus charge because of this oscillation. The complex space time oscillation causes equal charges to repel and unequal charges to attract.

The net result of these inter-dimensional space time oscillations cause the wave equations in physics. These equations were hard to understand as to why they exist. Yet the dual light speed dot-waves make it easier to understand at an engineering level.

Section 15-2: The Magnetic Fields

If we take an electron and move it through space, the electric field surrounding the electron will start to move and rotate. The dot-waves within space will align to the electron as the electron passes a particular point. Linear motion will cause the electron to spin as it moves. The vibrating dot-waves within space will tend to follow along with the electron and to spin as well.

The photon energy added to the electron will cause the dot-waves within space to move and spin. While this is happening the dot-waves are oscillating between the Co universe and the Cs universe. Thus the magnetic field exists within both the Co dimension and the Cs dimension.

Chapter 16: Conservation of Intelligence of Universe

16-0 Introduction

The universe is a perpetual oscillator. It oscillates from a minimum radius to a maximum radius forever. The oscillation always existed and will always exist. The universe always conserves energy as well. The energy goes from a Co/Cs oscillation to a Cs/Co oscillation. The energy is concentrated in a small volume and later in a large volume.

At the same time the intelligence of the universe remains constant. Intelligence is conserved. The intelligence goes from the contracting Cs/Co universe into the expanding Co/Cs universe. The intelligence that we generate is not lost. It remains within the radiated Cs/Co photons. The information necessary to produce man and life is stored in memory so that man and life will always return to existence. Then man will study the process by which man was created.

When we wonder how man appeared upon planet earth we look at scientific, philosophical and religious possibilities. The question of how the universe was initially created only exists when time exists. Once we eliminate time, there was neither beginning nor end of the universe. It always existed.

The universe is a perpetual oscillator. It grinds out stars and planets and life forms. The production of stars will cause heavy atoms to be produced. The ingredients of life will always be produced somewhere in the universe. Planets like Earth will always be produced.

Higher life forms such as man will always be produced. There most likely are millions of Earths in the universe. Some have man. Some have other highly intelligent creatures. Any creature of high intelligence that does not look like man can be considered by us to be an alien.

The result is that the universe is full of aliens. The physics of the light speed Co/Cs universe is such that the aliens are limited in their ability to travel the universe. We can dream of going faster than the speed of light Co but our spaceships will be destroyed if we go much beyond a speed of 0.2Co. The same is true of any Co/Cs aliens who want to visit us.

The multi-light-speed universe has unlimited possibilities. Their spaceships could travel much faster than light speed Co and could reach us. On the other hand, their universe would tend to be much further away from the big bang. Thus it is most likely that they would live in an entirely separated universe from us.

As we go to extremely high light speed universes upward toward light speed infinity, a point would be reached where no particles exist. Such universes would be purely photonic universes. This would lend itself to many philosophical and religious possibilities. This I discussed in my various other books over the years. For this book only the scientific solution is discussed.

For the situation where only the dual light speed Co/Cs universe exists, we are limited to only one type of alien that could encounter us. These aliens are low mass high light speed entities. Let us look at that possibility.

Section 16-1 The Intelligence of the Earth

The prior universe that we came from was the contracting Cs/Co universe. This compressed at a final velocity that is the geometric mean between light speed Cs and light speed Co. The big bang produced the expanding Co/Cs universe that we live in. A contracting universe tends to produce spherical shells of contracting energy.

Within these shell of energy is the intelligence required to produce life and intelligent creatures. In the contracting universe there is little room for individual life forms. Thus in the reverse side of the universe only collective intelligent structures existed.

The big bang enabled the production of particles and sub-particles and individual forms of life. At the same time there were some remaining Cs/Co forms of energy which produced photonic energy intelligent structures.

The stars generate a lot of radiated Cs/Co energy. However it is not conducive to the creation of a photon light speed Cs/Co mind. The center of ordinary planets also generates a lot of Cs/Co energy. This starts to form an intelligent structure.

The result is that the structure of a photonic mind reforms below the surface of our Earth and heading toward the center of our Earth. Within this mind is the data for the production of life and higher existence.

At the same time that the Mind of the Earth is forming, the Darwinian evolutionary process starts to develop. There is a feedback of intelligence between the light speed Co dimension and the light speed Cs dimension.

Man upon this Earth evolved to be man out of the intelligence which existed in the contracting Cs/Co universe. The Darwinian process is the crude power process. The feedback process is the fine power process. This is similar to a feedback amplifier in electrical theory.

The question of "Who made Man?" can only be answered that "Man made Man". Man always existed and will always exist. Man exists in the physical here and now and later man will exist as the memory of man.

As man grows in intelligence we learn the details of the Darwinian process. We learn the structure of life and how early life came into existence. The information that we learn returns to the universe so that the process will be repeated on the next cycle.

The process repeats upon millions of Earths. The loss of one Earth or many Earths does not eliminate the possibility of man coming to be. We are a very important ingredient of the life process. We learn and in so doing insure that the memory of the universe remains intact.

Section 16-2: The Dual Light Speed Mind of Man

We have a body and a mind in the light speed Co dimension. We also have a body and a mind in the light speed Cs dimension. We are dual light speed creatures. Most people do not realize that they exist in two different dimensions. Religious people tend to believe that they have a soul yet they do not know what a soul is.

Throughout the ages many people have communicated with something beyond their own physical being. Often people push their minds and bodies to the limit and enter a hallucinatory state where they see things and hear voices.

The questions are what is happening and who are these people talking to? To answer these questions requires an understanding of our dual light speed mind.

Our physical light speed Co mind is an electro-chemical mind. It is self-contained within our body and serves to keep us alive and continue the existence of the specie man. Our light speed Cs mind is an electro-photonic high speed mind. For the most part it mirrors our electro-chemical mind so that it is difficult for a person to tell that he or she has a dual mind.

When a person overstresses himself, the physical light speed Co mind can lose alignment with the light speed Cs electro-photonic mind. This will cause the person to hear things and see things. The person will believe that he or she is speaking to some spirit outside his own mind. Yet the truth is that he is merely communicating between his light speed Co mind and his light speed Cs mind.

Some people who are schizophrenic get caught in two worlds. One world is what his eyes see and his ears hear. The other world is a confusing world of collective intelligence. The Cs dimension ties our high speed mind to the Earth mind and the mind other people nearby. It is a horrible situation to be in.

The hyper manic bipolar mind is able to reach into the light speed Cs dimension and gain information from the Earth mind and others as well. Then the person can return to a normal state. He will have vivid dreams that bring him interesting ideas and data. It is a very creative mental handicap. When he is in a manic high state new ideas flow into his mind. The downside is that he is often unable to function normally during these periods.

The regular bipolar mind goes from a high or normal state to a very low depressed state. This is a terrible affliction with no real benefits.

The state of mind lock is a terrible condition where the person's mind is locked into his high light speed mind. He can be in a trancelike state for many years. Mental illness is often a chemical unbalance but it is also a situation where the physical light speed Co mind is not aligned to the light speed Cs mind.

Section 16-3: The Flow of Intelligence at Death

During life we create and absorb intelligence. Our intelligence is stored in our light speed Cs dimension. This takes the form of Cs/Co electro-photons. When we die our physical body perishes in the Co dimension and our light speed Cs body starts to die as well. At that point the intelligence we have acquired is transmitted to the mind of the Earth.

During near death experiences people have seen light. Their light speed Co mind has looked inside their light speed Cs mind and seen the mind of the Earth.

The mind of the Earth will process the data into memory. Thus the dead person will be reduced to memory. This will enable the possible reincarnation of the person into the mind of a new fetus.

Section 16-4: The Cosmic Reincarnation Process

Many people will reincarnate into new bodies upon this Earth. Information from the deceased with be absorbed by the mind of the Earth and transmitted into the mind of a forming fetus.

The young child will have this intelligence. It will be in his or her light speed Cs mind. Over time as the child learns and grows new data from his light speed Co existence will override much of the knowledge of his prior life.

The net effect is that the memory of a person extends beyond his present life into a portion of his next life. Then his present life slowly fades away and he becomes a new person.

The earthly reincarnation process is a flow of intelligence between the mind of the person and the mind of the Earth. Then the intelligence is purified and transmitted into a new fetus upon the earth.

The cosmic reincarnation process involves the transmission of data from one Earth to another. This data will flow at light speed Cs. This enables a person's intelligence to be transmitted huge distances from one Earth to a receiving Earth.

For the multi-light-speed possibility, the memory of a person could move upward toward higher and higher light speed levels of existence. This would enable a person to achieve a purely photonic type existence.

Short Biography of Author

I was born on Dec. 24, 1938 in Brooklyn N.Y. and I went to P.S. 25 in the Bedford Stuyvesant section where I received the Mathematics medal. Teachers used to call me little Einstein.

I went to Brooklyn Tech H.S. and graduated in 1956 when I lived in the Fort Green Projects across the park from the school. I took the electrical course and received the only 100 percent on the statewide Electrical comprehensive.

I went to the Polytechnic Institute of Brooklyn at night from 1956 to 1966 and I graduated B.S.E.E. (summa cum laude)

I lived in various locations in Brooklyn N.Y. while raising a Nephew since I was eighteen and he was three. I was like a working mom as I sent him to day care while going to work. I was lucky to find a young woman who was willing to take me and my nephew, and we got married in 1965 while I was finishing my education.

I worked at Con Edison as an Engineering Design Assistant from 1956 to 1962. There I designed the 4KV transmission lines and sub-station interconnections for the East Bronx section of N.Y. I also designed the 13KV distribution system for Freedom land Amusement Park which later became Co-Op City.

I worked for the City of N.Y. as an Assistant Engineer from 1962 to 1967. There I designed the lighting and modernized the electrical system on the Hutchinson River Parkway as the City went from incandescent lighting to mercury vapor lighting. I redesigned the lighting for the 59th street Bridge and I worked on many street lighting improvements and many small underpasses and, overpasses. I redesigned the Rockaway boardwalk lighting after a hurricane destroyed the old system.

I worked for Sperry Rand of Great Neck Long Island as an Associate Engineer from 1967 to 1970. There I worked on the old WDE military system and designed improvements to make it work better. I worked in both the test department and the Radar Research Department, and solved many problems on many systems. I left the job when the atmosphere deteriorated and many people were being laid off.

I worked for the Port of N.Y. Authority in Manhattan as an Engineer from 1970 to 1972. There I designed the new aircraft fuel pumping control system for J.F. Kennedy Airport. I worked on airport tower lighting for Kennedy and LaGuardia airports and I also checked and approved the underground lighting for the Lincoln center parking garage.

I worked for the Sperry Rand Radar Research Department as an Engineer from 1972 to 1993. Sperry was rehiring and I preferred to work there instead of the commute to the City since my house was in Plainview Long Island.

I designed the control system for the five inch guns for the Aegis class destroyers. The Navy called my system the "Maytag" because it worked the first time and every time. I worked on the sonar system and solved many problems on the Polaris missile system. I was the Engineer in charge of the first group outside of Univac to build their UYK 44 Univac computer and get it working. I worked with physicists on the ring laser gyro. I also worked on the Nexrad weather radar. When the company was Sperry Rand it was a great company. Things got bad when it became Unisys and huge layoffs occurred and the company downsized greatly. In 1993 the company was no longer doing any new and interesting work. They no longer had any need for their best engineers and there was a big layoff of 250 engineers in 1993 which gave me the opportunity to retire early at age 55.

When I retired in 1993 I went to work on The Dot-wave Theory and other Theories which had begun in 1981. I moved to Virginia Beach into a larger house with very low taxes as compared to Plainview N.Y. In addition I started a part time handyman business since I always enjoyed fixing things and painting. I always found various assistants to help with my small business. My reduced pension, my shrinking 401K, my small business and the profit on the N.Y. house held me over until my Social Security came in. With reduced income and property taxes, I found out that my retirement take home pay was the same as when I worked.

I was then free to work on my Dot-wave theory which I had started in 1981. I then self-published many books and sold some on the internet and gave most away. This was expensive as McNaughton & Gun charged over $5,000 to print 1000 to 1500 books.

As I moved from New York to Virginia Beach in 1993 I made a nice profit on the house which helped me over the years. When I moved from Virginia Beach to Cary N.C. in 2008, I also made a nice profit on the house. Finally when I moved from Cary N.C. to Sanford N.C. in 2014 I lost a little money on the house. The extra money always helped with the books and other expenses.

Times changed since Create Space and Kindle allows me to self-publish my latest books at basically zero cost. This is very helpful since I no longer have profits on the houses to pay for the books and other expenses. These last few years I rented a small single wide from relatives for low rent as long as I maintain and improve it.

I used my handyman talents to add modern bathrooms and cabinets. My place is small but I have a nice screened in front porch, an enclosed rear porch and a large metal shed which I built.

In any event I have given a little background of myself to give the reader a little understanding of my life and my effort in the production of my books. I am a singular researcher. All I have at my disposal is my mind, a fifteen dollar calculator, pen, and paper plus a modern home computer. This is better than the 1898 Underwood typewriter which I used for my early work from 1981 until it wore out from about ten thousand pages of typing when I explored a huge amount of ideas.

My work on the dot-wave theory started in 1981. Most of the basic equations in this book were written from 1981 to 1983 in my notebooks. I used to live in nice size houses and had plenty of storage space for my notebooks. However in the last few years I live in small rented quarters and have very little storage space. Therefore all my original notebooks are gone.

A lot of my work is in the copyright office under various titles. However the notebooks are gone. All that original work was for light speed C dot-waves with coexisting multi-light-speed universes going up to light speed infinity.

I finished my novel "Futureoids and Cosmic Reincarnation" in Jan 2018. Up until that time I believed in coexisting dimensions with a degree of interactions. Then it became self-evident there was strong interactions between our dimension and a very high light speed dimension. I then wrote the last three books on the dual light speed universe.

In the late winter I started to build a new shed for my daughter. The carpenters put up the basic outside and roof and I painted it and installed insulation and plywood inner walls. Then I added electricity, air conditioning, and television lines. I enjoy the work and glad that the hard work of the basic structure and roof were done by the carpenters. I still can do handyman work but at 79.5 years of age, I do not have the strength that I used to have when I was younger.

I thought that I was finished with my work until I watched a television program on quantum entanglement who indicated a light speed great than 10,000 C.

This caused me to restudy my original mass to charge velocity conversions and it became self-evident to me that the ratio of the gravitational constant to the permeability constant was the ratio of the high light speed Cs to our light speed Co. This information necessitated the writing of this book which I started in May 2018 and finished by July 2018.

At Parting:

This latest study is now at an end. It took about two months to rewrite my prior book. Today I was wondering why my inner mind defined the dot-waves as dots in 1981. This was at the start of my lifetime study. I did not know very much at that time other than what I had been taught in school.

If we look at the junction between the Co and Cs dimensions, the distance across the junction is very small. In addition the diameter of the worm hole between the dimensions is also very small.

The dot-waves will oscillate from the pinpoint to a maximum size in both dimensions. I did not realize that long ago. Yet my inner mind understood that it existed across the barrier.

It seems to me now that at the junction the plus dot-wave looks like a plus dot and the minus dot-wave looks like a negative dot.

Anyway thanks for reading my book.

Appendix: Standard Government values and calculated values

STANDARD GOVERNMENT VALUES

Quantity	Value
R_{BOHR}	5.29177E-11
C	2.99792E8
$(F.C.)^{-1}$	137.036
G	6.67260E-11
e_o	8.85418E-12
h	6.62608E-34
K	8.98756E9
Mile	1.609 km
M_N	1.67493E-27Kg
M_N	939.564MEV
M_P	1.67262E-27Kg
M_P	938.272MEV
M_E	0.910939E-30Kg
M_E	0.510999MEV
Q	1.60218E-19
Uo	1.25664E-6
Zo	376.731

CALCULATED VALUES

Quantity	Value	Equation #
Bohr velocity V_B*	1.21667E-28	5-57
Cycle Time Tu	1088Byrs	6-24
Dot-wave Charge Q_D	3.47119E-60	5-87
Dot-wave Period	1088Byrs	6-24
G	6.67223E-11	5-62
Radius Universe	1.30392E26	5-65
Mass of Dot-wave	1.96867E-71	5-83
Mass Universe	2.65773E52	5-73
Time Universe	4.34940E17sec	5-60
Time Universe	13.7827byrs	5-61
#Dot-waves electron	4.62718E40	5-85
#Dot-waves neutron	8.50793E43	5-92
#dot-waves proton	8.49619E43	5-88
#Dot-waves Red Photon	1.71063E35	11-5
#Dot-waves Universe	1.35001E123	5-94
#Neutrons Universe	1.586768E79	5-72
R_{BOHR}	5.29178E-11	5-59
Zo	376.729	5-38

Index of selected topics

Blank Note Page for Reader